E £ 30 -00
6B

VEGETABLE SEED PRODUCTION

VEGETABLE SEED PRODUCTION

Raymond A T George   University of Bath

Longman   *London and New York*

Longman Group Limited
Longman House, Burnt Mill, Harlow
Essex CM20 2JE, England
Associated companies throughout the world

Published in the United States of America
by Longman Inc., New York

First published 1985

British Library Cataloguing in Publication Data

George, Raymond A. T.
  Vegetable seed production.
  1. Seed industry and trade
  I. Title
  635'.0421   SB117
  ISBN 0-582-46090-5

Library of Congress Cataloging in Publication Data

George, Raymond A. T.
  Vegetable seed production.

  Bibliography: p.
  Includes index.
  1. Vegetables--Seed.   I. Title.
  SB324.75.G45 1985   635'.0421   84-12604
  ISBN 0-582-46090-5

Set in 10/11pt Bembo Roman, Linotron 202.
Printed in Great Britain at The Pitman Press, Bath

CONTENTS

PREFACE

The worldwide interest in vegetable crop production is increasing and because many of the species are produced from seeds there is more activity in vegetable seed technology. There are many different species cultivated as vegetables and it was therefore thought necessary to produce a book dealing entirely with vegetable seed production.

The present volume attempts to bring together the diverse information relating to seed production of the most important groups of vegetables. The first six chapters deal with basic principles of vegetable seed production and many of the concepts can be used throughout the world. The production methods and associated information relating to individual species of the widely grown vegetables propagated from seed are discussed according to their botanical families in Chapters 7 to 16.

The book is largely based on information required by the author when involved in either vegetable seed programmes in developing countries or teaching undergraduate and postgraduate students in the United Kingdom. It is therefore designed for use by students and those engaged in vegetable seed programmes in temperate and tropical regions of the world. The expert vegetable seed producer in commerce has a very deep knowledge of the techniques relating to the species with which he specializes, and although this book does not pretend to surpass this wealth of experience it is hoped that the seed production specialist will find it a useful reference when called upon to deal with unfamiliar crops. Each chapter has references directly relating to the text and a further reading list which can be used to obtain information on related topics.

It is my pleasure to thank vegetable seed producers and vegetable growers throughout the world with whom I have come into contact and I acknowledge the information and ideas gained from visits and encounters with farmers, technologists and research workers.

I am especially grateful to Dr Walther P. Feistritzer and other colleagues of the Seed Service, Plant Production and Protection Division of the United Nations Food and Agriculture Organization for their lively interest during the preparation of this text.

Above all, I express my great gratitude to Professor Leonard Broadbent who first suggested that I should write this book and who subsequently read the final drafts. His editorial and professional skills have made a very significant contribution to my original manuscript.

March 1984   Bath

ACKNOWLEDGEMENTS

We are indebted to the following for permission to reproduce copyright material:

The British Society for the Promotion of Vegetable Research for figs 13.3, 13.4 from figs 3, 2 p 165 (Gray 1981), & 13.5 from figs 1, 2 pp 171/173 (Gray 1979); Food & Agriculture Organization of the United Nations for table 1.3 from table 1 p 3 (FAO 1981); the author, Professor J Harrington for table 5.1 from table 2 p 92 (Harrington 1959); Inter-National Board for Plant Genetic Resources for table 11.1 from table 10 p 66 (Grubben 1977); the authors, Drs R Jacobsohn & D Globerson for table 13.1 from table 42.11 p 644 (Jacobsohn & Globerson 1980); Longman Group Ltd for fig 3.1 from fig 10.11 p 308 (Webster & Wilson 1980); National Institute of Agricultural Botany for tables 1.4, 1.5 from tables I, IV pp 48/51 (Bowring 1970); the author, Professor Rick for fig 12.4 from diagrams (c), (d) p 69 vol 239 (2) (Rick 1978); the author, Dr I Rylski for table 12.1 from table 4 p 151 (Rylski 1973); the author, Dr N H Sandin for fig 31.2 from fig 36.1 p 554 (Sandin 1980); Vegetable Research Trust for fig 3.4 from fig 1 p 17 (Faulkner 1983).

We have been unable to trace the copyright holders of fig 8.4 (Thompson 1938), fig 14.7 (from Naamni et al 1980), table 5.2 (after Delouche et al 1973), table 5.3 (after James 1967), and would appreciate any information that would enable us to do so.

1 ORGANIZATION

THE ROLE OF VEGETABLES

Vegetable crop production is one of man's basic skills. Wherever he has set-
tled for long enough to produce a crop he has cultivated vegetables for human
and animal food. The level of success and productivity originally depended
on the local climate and seasons, and the range of species cultivated. These
had been developed by selection from local wild plants and were supplemented subsequently by plant introductions from other areas and later still
from other continents.

It is now becoming increasingly appreciated that successful vegetable production is very dependent upon a supply of satisfactory seeds. At the present
time the seed industry plays an important role in both production and distribution of vegetable seeds. It must be emphasized that the purpose of this
book is to discuss vegetable seed which is used for reproducing the next generation of a crop rather than seed which is used as grain, although in many
instances the seeds of grain or of vegetable species such as dried peas or beans
are grown for food. But even in these instances there are often important
differences between the culture of seed which is eaten and that which is kept
for producing the next generation of plants.

The production areas of vegetables range from large-scale farm enterprises
and market gardens growing for profit to private gardens or homesteads,
where vegetables are an essential element of the families' own efforts to supplement their diet or income. Vegetables are also cultivated in some communities as physical recreation or even for a pastime or hobby. There are
some relatively small-scale producers who aim at self-sufficiency in vegetables plus a surplus for sale or exchange in village communities. The market
growers in many areas have evolved from this type of disposal of surplus
crops to deliberate production of crops for sale. With the further extension
and development of urban communities the commercial producer has continued to play an increasingly important role in meeting the vegetable requirements of the population. Commercial production has extended
considerably during the last few decades in many parts of the world as large-
scale enterprises endeavour to provide continuity of supply for the fresh
market, processors and export.

RANGE OF VEGETABLES

The range of types of vegetable is very diverse and includes staple root crops such as potatoes and yams, leaf vegetables such as *Amaranthus* and lettuce, bulb crops such as onions and garlic, and edible fruits, e.g. tomatoes and melons. Many of the legumes, e.g. *Phaseolus* spp. and *Cajanus cajan*, are not ranked as vegetable crops when the seeds are dried, but they are when immature pods or seeds are eaten, e.g. green beans (*Phaseolus vulgaris*) or immature pea seeds (*Pisum sativum*). When these types of crop are grown for large-scale production of the dried seed for processing or consumption they are usually excluded from vegetable crop statistics and defined as grain legumes or pulses.

CLASSIFICATION OF VEGETABLES

There are several systems of classifying the diverse crops grown throughout the world as vegetables:

1. According to their use or part of plant eaten.
2. According to season or area of production.
3. According to their botanical classification.

A classification according to their use is probably the best way of drawing attention to the diversity of species cultivated as vegetables, but we do not gain any other useful information regarding their cultural or physiological requirements (Table 1.1).

Table 1.1   Examples of vegetables classified according to part of plant used as a vegetable

Part consumed	Scientific name	Common name
Seedling	*Glycine max* (L.) Merr	Soya bean
Shoot	*Asparagus officinalis* L.	Asparagus
Leaf	*Amaranthus cruentus* L.	African spinach
Bud	*Brassica oleracea* L. var. *gemmifera* Zenk	Brussels sprout
Root	*Daucus carota* L. subsp. *sativus* (Hoffm) Thell.	Carrot
Bulb	*Allium cepa* L.	Onion
Flower	*Cynara scolymus* L.	Globe artichoke
Fruit	*Lycopersicon esculentum* Mill.	Tomato
Seed	*Phaseolus vulgaris* L.	Bean

When vegetables are classified according to their season of growth or climatic area, the diversity of species is apparent and there is some indication of each species' environmental requirements although some ambiguities occur. Also this can indicate the type of climate where seed production may or may not be successful (Table 1.2).

The most useful system to classify vegetable crops is probably based on

Table 1.2 Examples of classification of vegetables according to season

Scientific name	Common name
Warm season crops	
Hibiscus esculentus L.	Okra
Zea mays L.	Sweet corn
Cool season crops	
Apium graveolens L. var *dulce* (Mill.) DC.	Celery
Beta vulgaris L. subsp. *vulgaris*	Beetroot, red beet

their taxonomy or botanical families. This draws attention to important botanical points such as the method of pollination and vernalization requirement while at the same time agronomic aspects such as the need for crop rotation and specialized seed production techniques, e.g. harvesting method, seed extraction and cleaning are still emphasized.

The botanical or taxonomic classification system has therefore been adopted in the second part of this book for the majority of vegetable crops, with some of the less common but locally important vegetables dealt with in a miscellaneous crop chapter.

THE IMPORTANCE OF VEGETABLES

There are several reasons for growing vegetables, but the most important one is for food. Vegetables are essential in the diet, providing fibre, trace minerals, vitamins, folacin, carbohydrates and protein (Oomen and Grubben 1978; Hollingsworth 1981). There is increased interest, especially in developing countries, to supplement the staple – usually vegetable – foods with other vegetables produced locally.

Other reasons for an increasing interest in vegetable production include the use of leguminous crops in rotations to increase the soil's nitrogen status, use of vegetables as break crops in arable farming systems and the incentive to grow vegetables as high value crops for the fresh market, processing or export to earn foreign currency.

An important reason why the demand for vegetables has increased in many areas of the world is the development of improved and more diverse methods of vegetable preservation, such as canning, freezing and dehydration. This technology has led to a more diverse range of vegetable types and increased quantities being called for by the processors. Coupled with this there have also been rapid advances in the technology of vegetable production on a large scale, such as in Europe and the USA where rapid adoption of herbicides suitable for individual crops has led to crop establishment in a relatively weed-free environment and the subsequent use of precision drilling.

This increasing interest in vegetables, found in all scales of production, has led to local, national and international activity to improve vegetable seed supply and quality.

THE SEED INDUSTRY

There are two basic ways open for farmers to ensure a supply of seed for the production of the next generation of a vegetable crop. These are (1) saving seed from their own crop, or (2) buying in seed from elsewhere.

There are no figures available as to the percentage of vegetable crops in the world grown from farmers' own saved seed. In countries with well developed agricultural and horticultural industries, such as the UK, the use of farm-saved seed is negligible whereas in some developing countries as much as 90 per cent of vegetable seed requirements are saved by farmers and growers. The percentage of 'own saved' seed not only depends on the level of seed industry development in a country but for some crops it depends on the difficulty with which seed can be produced in the local environment and the amount of extra work a grower is prepared to undertake to produce his own seed. For example, it is relatively simple to save seed legumes but an extra season is normally needed to obtain seed from biennials such as onions. It can be argued that if vegetable growers produce their own seed, the plants for the seed crop have been subjected to selection pressures under the same environment as the market crop. It is for this reason that, until relatively recently, and despite a highly evolved seed industry, many Brussels sprouts growers in Bedfordshire (UK) seeded their own selected plants. There is still a place for local seed production of cultivars for specialist requirements, as is done by the winter cauliflower producers of Jersey (Channel Islands) and Cornwall (UK). But even in these examples it is probably better to produce stock seed locally and send it to a more suitable environment for the final stage of production. Also consortia of local growers, rather than individual growers, will be best able to ensure maintenance of genetic quality by satisfactory isolation from other cross-compatible species and cultivars.

In addition to management and economic factors associated with attempting to produce one's own vegetable seed, there are many difficulties resulting from the physiological requirements of some species in order to flower. For example, some of the Cruciferae do not flower in the tropics, where temperatures are never low enough for vernalization.

Vegetable seed production within a country may be a government or private venture. In some countries both government and private enterprise operate side by side.

In developing countries which do not have a vegetable seed industry the feasibility of producing vegetable seed is most likely to be instigated by government initiative and, as the appropriate infrastructure develops, there is a transition of responsibility to private enterprise. However, there is no hard and fast rule, the organization of responsibility and overall development rate will depend largely on priority given to vegetable crop improvement by individual governments. There is a general tendency for governments' interest in inputs aimed at vegetable crop improvement to follow in the wake of success with cereals or other appropriate staple crops.

There are several reasons why the initiative in vegetable seed production in a developing country is more likely to be taken by government than private enterprise. Firstly, there is unlikely to be any existing vegetable seed production expertise in commerce. Secondly, commercial companies within a country are unlikely to have the capital investment potential required for seed industry development; and even if they have, they are unlikely to be willing to invest in a commodity which requires the development of production units as well as the necessary buildings and equipment for harvesting, processing, storage and distribution.

The need for seed industry development in the developing countries calls for a development rate far in excess of the relatively slow rate at which existing seed industries have evolved in countries such as the UK and USA. An additional factor which sometimes calls for action by government rather than private enterprise in developing a seed industry is the country's socio-political beliefs and organization. A country which is isolated by its own or neighbouring countries' economic or social policies has a greater need to be self-sufficient.

Other factors which will have an effect on the rate of vegetable seed industry development include the emphasis put on vegetables and vegetable seed as a component of human nutrition or their importance as a means of earning foreign exchange.

In many countries there is already a well developed and effective vegetable seed industry. Countries such as the UK and the USA have developed their seed industries over many years, while countries such as India and Kenya have developed successful seed industries in a relatively short time with considerable consolidation and development of infrastructure, production and marketing capabilities.

The Food and Agriculture Organization of the United Nations (FAO 1981) has reviewed the extent to which seed production has been developed in individual countries. The countries are classified according to their geographical distribution and the level of vegetable seed activity, i.e. category A – advanced level; category B – fragmentary or pilot scale operation; and category C – no activity reported. The information is shown in Table 1.3 and clearly indicates that whereas some European and North American countries have well developed vegetable seed production and quality controls, there are many countries, especially in Africa, Asia, Central and South America, which do not.

Seed industry development

As the agriculture in a country develops, its government usually seeks to improve the vegetable seed situation. This often calls for financial and technical assistance.

There are many organizations which assist countries in the development of their seed industries. Some are capable of supplying both financial and technical aid, e.g. the United Nations, while others are either only in a position or willing to provide technical assistance for a seed project, rather than finance its implementation. Conversely, there are organizations with finance available for sound projects but which do not specialize in technical assistance and advice.

Table 1.3 World situation of production and distribution of quality vegetable seed 1979/80 (after FAO, 1981).

Region	Number of countries reviewed	Category A			Category B			Category C		
		Cultivar improvement (%)	Seed quality control (%)	Seed production and distribution (%)	Cultivar improvement (%)	Seed quality control (%)	Seed production and distribution (%)	Cultivar improvement (%)	Seed quality control (%)	Seed production and distribution (%)
Africa	25	0	0	0	44	36	40	56	64	60
Asia	24	0	4	4	79	50	54	21	46	42
Central America	9	0	0	0	44	56	22	56	44	78
South America	10	40	20	0	30	70	70	30	10	30
North America	2	100	100	100	0	0	0	0	0	0
Europe	11	82	91	82	0	0	0	18	9	18
Oceania	3	33	33	33	0	0	33	67	67	33
Total	84	19	19	16	44	39	39	37	42	45

Categories of national seed programme development level:
Category A = Advanced level.
Category B = Fragmentary or pilot-scale operation.
Category C = No activity reported.

The role of seed in agricultural and horticultural development

Most vegetable crops are grown from seed. It is important therefore that growers, economists, planners, administrators and occasionally even politicians appreciate that seed is the starting point for vegetable crop production. Other inputs or investments include items such as labour, fertilizers, irrigation, mechanization and crop protection (including fungicides, insecticides, acaricides, nematocides and herbicides). Many of these inputs depend directly or indirectly on fossil fuels and it is increasingly important that none of these important or valuable inputs is lost or wasted as a result of low quality seed.

The value of vegetable seed as an input expressed as a percentage of the total value of all other inputs to produce a crop is remarkably low. For example, for a field crop of direct drilled carrots it is approximately 20 per cent of the total cost; for a transplanted crop of cauliflower it is approximately 7 per cent of the total cost. In protected cropping the difference is even greater: the seed is less than 1 per cent of the total production cost of an early tomato crop produced in heated greenhouses.

Seed can be identified therefore as the primary and essential starting point of vegetable production programmes. This concept holds good for any scale of production whether it be in a small private garden or a large area for processing.

The role of seed industries in agricultural development

In many food programmes the quantity and quality of seed available will have a direct effect on availability of fresh vegetables in the markets. The range of types and availability of each of them over as long a season as possible will be affected by availability of seed of suitable cultivars.

New cultivars which are either developed within a country or introduced from elsewhere have to be distributed via the mechanism for seed supply. Many modern cultivars are relatively high yielding compared with their predecessors. There are cultivars which have been produced with resistance to specific pests or pathogens, for example tomato cultivars with resistance to specific nematodes, *Fusarium* and *Verticillium* wilts. Some have been developed by plant breeders for specific seasons or purposes, for example lettuce suitable for greenhouse production during the winter or 'stringless' dwarf beans for processing. Other cultivars, such as F_1 hybrid cabbages, have been produced to increase yield and have a higher degree of uniformity than older cultivars.

New vegetable crops to widen the range of produce available are introduced via the seed marketing or distribution network within a country or state. In this way new crops (for example, Chinese cabbage introduced into the protected cropping system of Northern Europe) or new methods with old crops (such as processed peas, frozen peas, canned tomatoes, tomato paste, onions for dehydration and cabbage for cole slaw) have been introduced into the horticulutural industry.

Some of the technological developments made by agricultural and horticultural research institutes have been adopted by seed companies. Examples include seed treatments for control of specific diseases, e.g. the hot-water treatment of celery seed to control celery leaf spot (*Septoria apii*). These treatments are done by seed companies before the seed leaves them. This technique has been largely superseded by a Thiram suspension but is still done by seed companies when requested before seed is dispatched. Similarly, grading and pelleting of seed-lots to facilitate their use in conjunction with precision drills or blocking machines for plant propagation is done for growers by the seed trade.

The possibilities of including crop protection chemicals, growth regulators, materials to accelerate germination under dry conditions and nutrients in the materials used for pelleting seed are constantly under review. Any positive steps as research progress is made will be taken up by seed companies for the benefit of growers.

In addition to the role of vegetables in improving the standard of living by way of diet and health, vegetable crops are a part of agricultural systems. Any improvement in agriculture's income and contribution to profitability from high value horticultural crops such as vegetable and fruit crops will play an important role in improving a country's economy.

There are therefore many indirect and direct ways in which a dynamic vegetable seed industry can assist a country, regardless of the stage of development that the country has reached.

SOURCES OF FINANCIAL ASSISTANCE FOR SEED INDUSTRY DEVELOPMENT

There are several sources of financial assistance for seed programmes in developing countries and the attitude of the donor or lending organization may depend on the level of agricultural development or political implications of the loan. The following types of organization may be involved in financial assistance for seed industry development.

The United Nations agencies

The United Nations (UN) generally finances seed programmes via the United Nations Development Programme (UNDP). Occasionally special funding may be available from the UN Food and Agriculture Organization (FAO). This is sometimes made available from FAO's special fund for Technical Co-operation Programme (TCP). A major role in seed programmes within the UN system is played by the FAO Seed Improvement and Development Programme (SIDP).

FAO Seed Improvement and Development Programme (SIDP)

This programme described by Feistritzer (1981) provides a range of assistance for seed programmes in developing countries. The SIDP is based in FAO

Headquarters in Rome but liaises with governments, international centres and organizations such as The International Seed Testing Association (ISTA). It meets requests for assessment of seed situations in developing countries and assists in formulation and implementation of programmes. The SIDP is able to draw on expertise from a wide range of seed specialists in the world and can act as an agency, within the framework of FAO, to locate specialists for specific assignments related to seed programmes and in some instances can assist in location of suitable donors for project funding.

Development banks

These include the World Bank, The Asian Bank and The Arab Development Bank. Generally finance for projects is loaned at fixed rates of interest and repayment is made over a relatively long term.

Bloc aid

This type of funding can be within political groups of countries or from a group of countries formed for trading and other purposes. Examples of such groups giving aid for seed projects include the European Economic Community (EEC), the Near East Governments' Co-operative Programme and the Arab Agricultural Development Organization.

Bilateral aid

This is the direct financial assistance from one country to another. It is not necessarily between adjacent countries, for example a European country may well give bilateral aid to an Asian country, e.g. UK to India or West Germany to Sri Lanka.

Other forms of bilateral aid may be provided by an official organization of the donor country. Examples include the Swedish International Development Agency (SIDA) and The Danish International Development Agency (DANIDA) both of which are actively supporting overseas seed programmes.

Other forms of financial assistance or aid may be given by organizations such as the United States Agency for International Development (US AID), the Ford and Rockefeller Foundations.

Self-financing

Finally the possibility of a country partly or totally supporting its own seed development programme remains. This can be at national, state or departmental level.

The extent to which a country is prepared to provide financial assistance for its own seed industry development is often seen by donor or loan agencies as a measure of the country's intent. Whereas financial inputs from outside the country are sometimes the only way of getting a proposed scheme started, it is the host country which must maintain and further consolidate the infrastructure once the initial project has come to an end.

SOURCES OF TECHNICAL ASSISTANCE FOR SEED INDUSTRY DEVELOPMENT

The initial teaching, training and technical advice relating to the development of a seed industry is usually provided by specialists from countries which have already achieved a satisfactory seed industry status. The specialists are either in full-time employment with organizations which provide assistance or are released from their normal duties on short-, medium- or long-term loan. Thus specialists can be made available from national government departments such as ministries of agriculture or educational establishments such as universities. Assistance of this type is often seen by the donor country as an important means of assisting developing countries. It is usual practice for the specialist to work with counterpart personnel in the recipient country. The availability of local counterparts is vital because the general concept is that when the development project has been completed, say after a period of three years, the local counterparts will continue more or less unaided having in the meantime become more experienced and conversant in the required technology. This transfer of information and development of skills at all levels is frequently referred to as 'transfer of technology'. In addition to the different national and international agencies mentioned earlier, there are consultant organizations which employ their own specialist teams and are frequently engaged by a government or other organization on behalf of a developing country to organize and execute a specific development project. These consultant organizations frequently compete with each other to secure a specific contract by submitting tenders with outline programmes.

ASSESSMENT OF SEED PROGRAMME REQUIREMENTS

An assessment of the vegetable seed situation in a specific country is essential before recommendation for seed industry development can be made. It is therefore very important to obtain information, data and records of observations from a wide area and to consult as many sources as possible in the search for information.

The assessment should study first the general vegetable crop situation to determine the constraints which result directly or indirectly from insufficient seed of satisfactory quality. It should include observations on agronomic practices such as sowing rates, cultivar suitability and presence of seed-borne pests and pathogens.

Having obtained an overview of the current vegetable production, the different facets of the seed industry and supply should be evaluated; all types of seed sources should be included, for example governmental, commercial and growers' own saved seed.

The points which should be included in the evaluation are summarized below. A systematic procedure will ensure that sufficient background information is obtained so that recommendations for comprehensive vegetable seed projects or for the improvement of any part of vegetable seed production, supply or quality control can be based on the findings of the appropriate parts of this framework.

1. *Background*   Relation of vegetable crop production to overall programmes of agricultural development. Relation of vegetable seed programme to overall seed programme.

2. *Vegetable crops produced*   Their location, relative importance, market trends; possible changes in emphasis or importance of specific crops; interaction of vegetable crops with other agricultural crop or animal husbandry groups; effectiveness of research or extension services on improved technology; sowing rates and estimated seed requirements.

3. *Cultivars*   The range of cultivars available, their suitability for local environmental conditions, production systems and market requirements; origins of cultivars (i.e. local or imported), rate of replacement; existence, scope and efficiency of cultivar trials.

4. *Seed Production*   The seed industry status and the vegetable seed production relative position and development. Quality and development stage of any existing vegetable seed production. Sources of vegetable seed used by growers. Identification of existing or potential vegetable seed production areas.

Level of husbandry and technology used in current vegetable seed production; current seed yields; existence of distinct seed classes.

5. *Seed processing*   Level of development of seed processing and degree of mechanization for harvesting, threshing and cleaning.

6. *Seed storage*   Adequate storage technology in use; effects of local conditions on vegetable seed quality and quantity. Adequate reserves for contingencies. Effectiveness of stock control, records of history of separate seed-lots and efficient labelling system.

7. *Seed marketing*   The extent to which the current vegetable seed is meeting demands of all areas and types of production. Adequate promotion of vegetable seed and range available, role of extension services and demonstration of improved vegetable seed.

8. *Legislation*   Existing legislation for other crop groups; inclusion of vegetable seed in existing schemes for quality control, germination or other such testing services or certification schemes.

9. *Policy*   Existing national and/or local government policy affecting vegetable seed requirements; adequate finance and manpower available for estab-

lishment or further development of requirements. Existence of training schemes for all levels of personnel including technicians.

SEED AS AN INTERNATIONAL COMMODITY

Since the development of specialist seed production areas in different parts of the world there has been a tendency for companies to specialize in the multiplication and distribution of seed. The actual quantity of seed produced will depend on several factors including the cultivar's multiplication rate (largely genetically controlled), seasonal conditions, husbandry and skills of managers, growers and technicians. Many of these companies are based in countries such as the UK, The Netherlands and France. A large proportion of the seed handled by such companies is grown abroad, often in another part of the world. A seed company based in the UK, for example, is likely to be producing the same species, even the same cultivar, in different areas of the world simultaneously. Pea seeds, for example, may be produced in Hungary and New Zealand by the same company using different contractors. A company will usually supply stock or basic seed to an overseas farmer (or contractor) on the basis of buying back from him all that is produced, but paying him according to the quality and purity of the seed-lot produced.

Seed companies continue to look for new production areas and new markets where not only climatic conditions but also the economic situation, including transfer of capital investment and local labour costs, are favourable. In this way there has been increased activity by seed companies to establish production contracts in countries such as Indonesia, Taiwan and Pakistan.

At one time seed of some crops was produced in commercial quantities on the off-chance that it would be purchased by an entrepreneur and subsequently re-sold at a profit to other people in the distribution chain. There is considerably less of this type of activity now, especially with vegetable seed, and most of the production is planned according to short- or long-term stock requirements, based on a specialist knowledge of the trade. Despite this stabilization, seed companies continue to look for new markets where sales potential is likely to increase (for example the development of large-scale vegetable production in a developing country).

This stabilization and organization of the seed industry has probably been a major issue in the advancement and introduction of legislation leading to quality control.

All this international activity and trade means that seed can be subject to price fluctuations or inflation resulting from monetary changes, for example international exchange rates and premiums on foreign exchange. The existence of tariffs or similar trade barriers will affect price and possibly the production locations. Other economic hazards or factors which can have an effect include trade embargoes or sanctions such as the UK seed trade experienced during the period when trading with Rhodesia was discontinued. Political or religious issues also come into play; Israel, for example, has a good climate and level of seed production expertise but cannot directly export seed to neighbouring Arab countries, which import vegetable seeds from seed houses further afield.

The International Seed Trade Federation

The International Seed Trade Federation (FIS) (Fédération International Semances) was formed in 1924 to discuss common problems and interests. At that stage of world seed industry development there was relatively little co-ordinated international activity. The Federation is composed mainly of national seed trade associations of some thirty or more countries plus individual companies as 'corresponding' members from countries which do not have seed trade associations.

In addition to dealing with business and trading problems with which the seed industry is faced from time to time, the FIS advises on the settlement of international disputes according to rules agreed and accepted by members. FIS has sets of rules for each of the main crop groups, i.e. herbage seeds, cereal seeds, forest tree seeds and vegetable seeds. The rules deal with such items or aspects of the international trade as contracts relating to offers and sales, import and export licences and definitions of trading terms. Seed quality is also defined in relation to trade, and the information required to be given with a seed-lot is listed. Other aspects included are business requirements such as shipping, insurance and packing agreements, and there is a formula for calculating compensations relating to contract disputes. The organization sees its own rules as being more realistic and appropriate to its arbitration procedures than international law, which can be time consuming and detrimental to satisfactory settlement within the life of a given seed stock.

The FIS co-operates with other international organizations involved with seed, including The International Seed Testing Association (ISTA) and The Organization for Economic Co-operation and Development (OECD). The more detailed functions of FIS were described by Leenders (1967).

THE ROLE OF PLANT BREEDERS IN SEED INDUSTRY DEVELOPMENT

The science of genetics and plant breeding has been applied to the maintenance and production of vegetable cultivars only since the beginning of the twentieth century. Even at the present time there are many cultivars, not only those of local importance, which are maintained by growers who simply save seed from selected or even unselected plants.

A good example of early cultivar development by people with an 'eye for a good plant' rather than scientific breeding is the diversity of cultivars of the *Brassica oleracea* subspecies which includes Brussels sprouts, cabbages and cauliflowers. These were produced and maintained mainly by individuals working in isolation and often with no biological training. This selection by gardeners and growers has resulted in the wide range of types within each subspecies. Examples are cabbages for successive seasons and with relatively different morphology; cauliflowers suitable for specific production areas and the different types of sprouting broccoli. The history and development of these cole crops was discussed by Nieuwhof (1969) who reviewed the important gardeners' and growers' skills which produced the present wealth of germplasm within them. Modern concepts of inbreeding and production of F_1 hybrids have been practised in certain cole crops only during the last twenty years or so.

The diversity within other vegetable crops can also be attributed to the selection process of successive generations of gardeners and commercial growers. The wide range of melons and other species in Cucurbitaceae are examples of this form of selection in arid and tropical areas of the world.

In more recent times the application of modern genetics and plant breeding has produced a rich record of vegetable crop improvements, including important yield increases. For example the modern yield of marketable Brussels sprouts shows a sixfold increase per unit area, and that of cauliflowers a fourfold increase achieved by improved cultivars introduced well within the last twenty years (Watts 1980). There are many other examples of improved vegetable crop production in the world resulting from the diligence of modern plant breeders and these include plant characters such as resistance to specific diseases, for example resistance to the wilt pathogens in tomatoes.

Plant breeding programmes can be initiated at governmental and commercial level, and in many countries they exist simultaneously in both spheres. For example, in the UK the Agricultural and Food Research Council (AFRC) supports the breeding and development of vegetable cultivars at the National Vegetable Research Station (NVRS). Similarly, in the Netherlands, the Institute of Horticultural Plant Breeding (IVT) (Instituut voor de veredeling van Tuinbouwgewassen) supported by the Dutch Government, works on cultivar development. Stations such as NVRS and IVT may work on plant breeding and genetical problems which would require too high an investment of time or money to be undertaken by private or commercial workers. In addition to the development of breeding techniques, government controlled stations usually demonstrate the application of a particular technique or breeding method by producing a new cultivar as the end-product. In some cases there is close co-operation between national institutes and private companies.

The emphasis of recent and current plant breeding programmes in the world is generally on the improvement and production of the staple food plants including cereals such as rice, corn (maize) and wheat; industrial crops such as cotton and sugar beet; and forage crops such as clovers and grasses. The total number of breeding programmes involving vegetables is considerably less than in these cereal, industrial and forage crops. Generally, improvement of a particular vegetable species depends on its local importance. The potato, *Solanum tuberosum*, is an exception to this cultivar development but it is still mainly propagated vegetatively.

There is international exchange of technical information, including breeding of new lines of the potato, sponsored, co-ordinated and executed by the International Potato Centre (CIP) in Peru with its seven regional offices in Colombia, Costa Rica, Kenya, Turkey, Pakistan, India and the Philippines. CIP also conducts specific country programmes currently in Tunisia, Nepal and Rwanda. There are also specialist national centres which include potato breeding in their overall programmes, for example the Plant Breeding Institute (PBI) at Cambridge, England. No other single vegetable species has such a high level of support for its improvement.

The tomato, *Lycopersicon esculentum*, has received a great deal of attention by plant breeders throughout the world and enormous progress has been made during the last few decades with programmes for breeding cultivars with specific characters suitable for greenhouse or field production systems, particularly in relation to disease resistance. But progress with the tomato has been generally achieved by individual institutes and associated workers rather than by concerted national and international effort. Notable success in cultivar development for greenhouse production has been attained by the Glasshouse Crops Research Institute (GCRI) in England and some Dutch commercial seed houses. A high level of success has been reached by American plant breeders in developing tomato cultivars for field-scale production using the determinate (bush) types.

The main objectives of plant breeders should be the improvement of yield and quality of vegetables with particular attention to pest and disease resistance. In addition, it is necessary to take into consideration the changes taking place in agronomy and marketing. These include production of cultivars tolerant to specific harvesting, processing or storage techniques. It is important that plant breeders are seen to be continually considering the requirements of the consumer and subjecting new lines to appropriate tests relative to the crop's market outlet.

It is only relatively recently that plant breeders working for European or North American vegetable seed companies have been deliberately aiming at overseas markets for their new cultivars. Some companies specializing in the export markets have probably been very aware of overseas trade when developing F_1 hybrids. Apart from any biological advantages of F_1 hybrids there always remains the commercial thought if a new or existing market outlet adopts F_1 hybrids, not only will growers be deterred from saving seed which would otherwise lead to a reduction in market potential, but the increased revenue received for F_1 hybrids will justify the cost of breeding, development and maintenance.

However, it must be emphasized that commercial plant breeders are generally working for the long-term needs of the industry rather than producing short-term novelties or cultivars with forms of disease resistance which will quickly succumb to adaptation pressure from pathogens.

The national and international interest in cultivar development has promoted attention to the need to collect and preserve germplasm. In the staple foods a clear priority can be identified, but with the diversity of genera and species utilized as vegetables throughout the world it is difficult to ensure that useful genetic resources are not lost. The International Board for Plant Genetic Resources (IBPGR) is active in identifying, describing and promoting national and international collections of germplasm of a range of species including the important vegetables in the world. Genera receiving attention in this way include *Allium* spp., *Amaranthus* spp., *Brassica* spp., *Capsicum* spp., *Lycopersicon* spp. and several genera in Cucurbitaceae. The IBPGR also sponsored the production of a report (Grubben 1977) which lists important vegetable species, records their economic and nutritional importance and location, and outlines proposals for their collection, conservation, evaluation and use as germplasm. A great deal of progress has been made in setting up the mechanism for this important activity related to future breeding programmes of crop plants including vegetables.

Plant breeders' rights

The granting of plant breeders' rights to the breeder of a specific new cultivar provides him, or his institute, with protection from other persons who could otherwise freely multiply or reproduce his material. The ease with which a cultivar can be reproduced by others in addition to the original breeder depends on the breeding method used in its development.

The reproduction of F_1 hybrids normally requires possession of the appropriate inbred parents, but it is relatively easy for other people to reproduce new cultivars which are predominantly self-pollinated once they have access to the material via normal commercial channels. Thus the registration of a new cultivar and the description of its distinct characters with a registration authority which is recognized in law protects the breeder from unlawful reproduction of his material by other persons in countries which have adopted this system.

In countries adopting this protection for plant breeders, the exclusive rights are given to breeders whose material has emerged from officially conducted field tests which have clearly shown it to be a 'new cultivar'.

Once plant breeders' rights have been granted, usually for an initial period of ten years, only the breeder or his appointed agents are legally able to reproduce the registered cultivar. This enables the breeder to recoup the cost of development in subsequent commercial sales of seed of the cultivar and also offers an incentive to encourage individual breeders or organizations to invest in the development of new cultivars. Before plant breeders' rights were established it was very easy for other interested parties to obtain another breeder's material once it was available on the market, reproduce it and offer it for sale, possibly even at a lower price than the original breeder could afford to do.

The system of operating plant breeders' rights in Europe is based on the International Convention for the Protection of New Varieties of Plants (UPOV 1974). This system is open to adoption by any country which agrees to ratify it. At the present time the majority of signatory countries are in Europe but it is visualized that the scheme will spread to more countries in other parts of the world.

The USA has a similar system for registration, but more emphasis is placed on the breeder's own description of his material rather than on field trials conducted by a registration authority.

Before a new cultivar can be entered on a European list of registered cultivars it must be grown and evaluated in offical trials, usually for a minimum of two successive seasons. This allows to some extent for reaction of the material to seasonal differences. During these trials the material is examined for distinctness, uniformity and stability (DUS). In addition to these DUS tests the cultivars submitted are evaluated for other important characters such as overall cropping performance and susceptibility to pathogens. The breeder submitting the material, or his agent, pays an agreed basic fee which contributes to the overall costs of the trials.

PLANT NAMES

The term variety has traditionally been the word given to an assemblage of the same species which have the same plant characters distinguishing them from other assemblages of the species. The rules of plant nomenclature were studied in the 1950s by a working party, and their initial considerations and code were later reviewed and revised (Gilmour 1969). The code is accepted by different countries as a basis for legislation relating to nomenclature. According to Article 10 of the International Code of Nomenclature, the terms cultivar and variety used in the sense of cultivated variety are taken as exact equivalents. Thus both words are currently in use in the industry. Many official titles of acts or legislation still prefer to use the term variety rather than cultivar. For example, the organization concerned with providing guidelines for conduct of distinctness, uniformity and stability tests on vegetable cultivars in The International Union for the Protection of New Plant Varieties (UPOV) and the UK Acts and Regulations still retain the word variety rather than cultivar, but many organizations and authorities are encouraging the use of cultivar in preference to variety.

All vegetables in cultivation are clearly named. Each has a generic name, a specific name and a cultivar name. For example, if we are referring to a certain cultivated variety (cultivar) of tomato known to growers as 'Moneymaker', it is identified as: *Lycopersicon esculentum* cv. 'Moneymaker'. In this example, *Lycopersicon* is the generic name, *esculentum* is the specific name (sometimes referred to by botanists as the specific epithet) and 'Moneymaker' is the cultivar name.

In order to avoid confusion with the naming of cultivated vegetables, especially new ones, there are clear rules of nomenclature, which have become increasingly important in countries with national trials and lists.

Cultivar names

The International Code of Botanical Nomenclature is not enforced or upheld internationally by legislation but many governments voluntarily subscribe to its success by adopting its recommendations, which are generally accepted by plant breeders. A plant breeder, or national breeding institute, will not normally give a cultivar name for a new line until the material has been seen to perform well in national cultivar trials and is likely to be acceptable on a national list. Until this is clear, the material is entered into trials under a code number or name. This avoids subsequent problems arising if the material is either not acceptable as distinct from, or has no clear advantages over, existing cultivars. Thus the proposed name, which may have a seed house's or breeding institute's prefix, is not associated with inadequate material. Similarly, if a breeder has a special name in mind, it is not wasted on what may be considered subsequently as deleterious material.

The International Code of Nomenclature lists cultivar names which should be avoided; the main points relating to vegetables and which must be excluded are:

1. Arbitrary succession of letters, abbreviations or numbers.

2. An initial article, unless it is the linguistic custom.
3. Names commencing with an abbreviation.
4. Name containing a form of address.
5. Names containing excessively long words or phrases.
6. Exaggeration of the qualities of the material.
7. Names which are likely to be attributed to other cultivars.
8. Names which may be confused with existing cultivar names.
9. Inclusion of words such as 'cross' or 'hybrid'.
10. Names exceeding three words (where an arbitrary sequence of letters, number or abbreviation is counted as one word).
11. Latinized names.

CULTIVAR RELEASE

Traditionally, growers, farmers and gardeners have become aware of new cultivars via the range of lists, catalogues and other media of the commercial seed companies. This practice still prevails for new vegetable cultivars in many countries. In others, however, legislation insists that cultivars of some crops are examined at national level to ensure that new cultivars are only released into commerce if they are clearly distinct from others already on the market. This information is also used before cultivars are selected for inclusion in trials for national lists of recommended cultivars.

There are several advantages in this system including avoiding synonyms, providing information necessary for the effective enforcement of plant breeders' rights, providing the consumer with information on new cultivars with their cropping performance and morphological characters.

There has been a concerted effort by interested parties, including the seed trade, breeders' and growers' organizations, to reduce the number of synonyms of vegetable crops. The existence of synonyms in lettuce types was demonstrated by Watts (1955), in spring cabbage by Johnson (1956), in red beet by Holland (1957) and in forcing radish by Watts and George (1958). There were two main reasons why synonyms such as those cited by these authors had come into being. Many synonyms had become popular because different lots of seed material from the same original stock were taken and multiplied up by different firms over successive plant generations. These separate lines remained sufficiently similar to be regarded as synonyms. Secondly, many seed companies preferred their own cultivar name for a particular genotype and were offering the same seed stock under different cultivar names even though they had each obtained it from the same primary supplier. This practice was not illegal at that time and the introduction of UK legislation to avoid this situation occurring is relatively recent. In practice, many customers of individual seed companies had preference for a seed stock with a cultivar name associated with that particular company.

With the introduction of plant breeders' rights and the formation of national lists it has become necessary to ensure that any new material released on the market is in fact clearly different from existing material. The distinct characters of such new material are used to distinguish the cultivar for registration under plant breeders' rights. This distinctness, plus proven uniformity of the cultivar and its stability for these characters when multiplied over several generations, is now examined and tested in official trials necessary for the 'lists of recommended cultivars' compiled by official organizations within an individual country, or occasionally groups of countries. Where several countries in close geographical proximity have trading or other agreements, there is now a tendency for them to compile group lists of cultivars for their community. The lists issued by the European Economic Community (EEC) are an example of this trend.

Distinctness, uniformity and stability tests

The evalution of cultivars for distinctness, uniformity and stability (DUS) enables controlling authorities to regulate the release of cultivars via national lists and control plant breeders' right (see above).

It is recognized that while a specific cultivar is a relatively unique genotype there is firstly the problem that the material may not be completely homozygous. For example, whereas an F_1 hybrid would be expected to be composed of a uniform population, an open-pollinated cultivar of a species which is largely cross-pollinated will display a degree of difference between individuals in the population. In addition successive multiplication will put different selection pressures on the cultivar. These selection pressures will depend on a range of factors including the pollination system of the species (i.e. whether predominantly self- or cross-pollinated), the climate and environment where the multiplication takes place, and the criteria used by the person selecting plants to produce the next generation. The DUS tests and their methodology used in the UK for testing vegetable cultivars were discussed by Webster (1974). The criteria used depend on the species but several characters such as morphology, time of flowering and resistance to specific pathogens are useful where they are less likely to be affected by environment and so are any characters which are clearly present in each individual plant of a population.

Guidelines for characters recommended for use in some specific crops were published by UPOV (1973). It is accepted that the general concept is one of assessing the identification of improved cultivars and accelerating their availability to the vegetable producers and industry. In the UK the DUS tests are conducted by the National Institute of Agricultural Botany (NIAB).

Cultivar trials

There are several reasons for conducting field trials for the evaluation of cultivars belonging to the same species. Trials which are organized specifically for cultivar registration or DUS testing have been discussed earlier. In addition plant breeders may sometimes conduct trials to examine all the available material as a prelude to a specific breeding programme. The results of this type of plant breeders' screening trials are not always published but a typical example was described by Fennell and Dowker (1979) who screened onion material for characteristics required for a breeding programme to improve autumn sown onions.

The type of cultivar trial discussed here is that which in broad terms assists farmers and growers to select from a list those cultivars which are suitable for specific seasons, production systems, maturity periods and market outlets. This type of trial is still frequently referred to as a 'variety trial', although the term 'cultivar trial' is increasing in popularity.

It is important that the objectives of a cultivar trial are clearly defined at the outset. All too often, especially in developing countries, the so-called 'variety trials' are simply a cursory look at the plants grown from free seed samples randomly and indiscriminately obtained from the seed trade. This type of 'screening' of available material can sometimes be useful as a 'growing out test' in a single season in order to obtain a general impression of material before putting it into proper trials, but they are clearly not cultivar trials.

A well organized cultivar trial should provide information for farmers and growers on the suitability of cultivars for specific purposes. If the trials indicate a wide range of deficiencies in the available material it may be necessary to search further afield for suitable cultivars, or alternatively to undertake subsequent breeding programmes.

Cultivar trials can provide information on where priorities should be placed in relation to national seed production or importation, and are essential in developing countries when deciding on a seed programme. Lists can be compiled from results of trials to help farmers, extension workers (advisers) and other interested parties. Kelly (in Feistritzer and Kelly 1978) defined three types of list which can result from cultivar trials: each contains 'preferred' cultivars (i.e. cultivars which are better than those not listed) and the lists are either *descriptive, recommended* or *restrictive.*

Descriptive cultivar lists

This type of list aims to assist farmers and growers to choose from the available cultivars, and would include those which have been shown in trials to be useful and acceptable. It would exclude vegetable cultivars with major faults within a crop group, such as low yield, frost susceptibility, poor fruit shape or poor storage quality. This type of list would enumerate the pros and cons of a particular cultivar so that a grower is able to decide which one is best for his particular needs. Information provided would be suitability for a specific soil type, season, market outlet and disease resistance. Separate lists may be published for different types of growers such as commercial producers and home gardeners. They are widely used in countries with well developed agricultural and horticultural industries and can play a major role in developing countries by drawing attention to suitable seed material which is available.

Recommended cultivar lists

This type of list is relatively short and sets out to advise growers which cultivars are firmly recommended for specific crops and purposes. The authority providing the information (such as a government) is not confining the growers' choice of cultivar to those on the list but clearly advising on the best material available.

Restrictive cultivar lists

This type of list has the objective of restricting the choice of cultivar to those on the list. Cultivars not on the list are prohibited and are not allowed to be offered for sale or to be grown.

Cultivars should not be restricted without good reason, and it is normal practice for a restrictive list to contain more cultivars of a specific crop than the recommended list which could also be issued. This allows for variations in availability due to seed yield fluctuations resulting from contingencies beyond the control of the grower.

Responsibility for organizing cultivar trials

The cultivar trials should be organized and conducted by an impartial agency. In practice, this is usually a government department such as a branch of a ministry of agriculture. In the ideal situation the work is done by an institute or organization which is government financed but separate from all other departments. In England and Wales, for example, the cultivar trials are conducted by the Trials Branch of the National Institute of Agricultural Botany (NIAB).

Individual countries or groups of countries have their own approach to this and usually delegate responsibility to organizations in existing infrastructures. The primary points to observe are that the trials must not be conducted by companies involved in the sale of seed or by plant breeders who have an interest in the trial results. In some countries the cultivar trials are conducted by university departments, agricultural and horticultural research stations or other relatively 'neutral' agencies. But in these cases the agency conducting the trials is operating under a contract from the country's ministry of agriculture or another government department with interests which are relatively close to agriculture and horticulture.

Organization of cultivar trials

Separate trials must be organized for each type of crop, even to the extent of conducting separate trials for specific market outlets. For example, if a trial is needed for the evaluation of outdoor tomato cultivars it is necessary to have separate experiments for the bush (determinate) and trained (indeterminate) types. But whether or not different production systems are included will be influenced by the production techniques used by growers in the area of the country where the trials are done. Similarly, season of cultivars is considered too; thus, for example, testing cultivars of bulb onions for their suitability to autumn sowing outdoors should be independent from a trial to test cultivars suitable for spring sowing.

Design of trials

The trials are usually based on a randomized block design with three replicates. The results are analysed statistically using an analysis of variance. Methods of analysing trial data were described and discussed in detail by Silvey (1978). Organizations responsible for trials usually have several sites in the country associated with the vegetable crop production areas.

In addition to the replication within individual sites, it is usually important to plant several different sites simultaneously. All available sites are not necessarily used for all crops. Most trial organizations have a 'home' ground adjacent to their headquarters and also have sub-stations in key areas. The sub-stations can be on satellite trial grounds, or sites such as university farms and experimental stations of ministries of agriculture. The use of sites rented from commercial growers is not usually recommended as it can result in problems relating to access, security or even agronomy.

The trials are normally conducted over at least a three-year period which allows for seasonal differences. It would be premature to make recommendations from successive trials in a shorter period. In practice the established trial organizations run each specific crop trial annually. A standard commercial cultivar which is well known to growers is used as the 'control'. This has two advantages: firstly, material familiar to growers is included and, secondly, the 'control' is a useful basis for comparing factors such as time of harvest and yield. The controls can be changed after several years if other consistently better cultivars are found from the results of the trials.

Allowance must be made to examine a range of harvest dates; for example, spring cabbage cultivars may be suitable for early harvest as 'greens', i.e. before harvesting, or may be grown on and harvested as 'hearted' cabbage. In order to assess suitability for both these stages of harvest a split plot design is used. Similarly, with Brussels sprouts, it may be considered necessary to examine single harvests but at a succession of four dates. In this case, eighty plants would be sufficient in the cultivar plot of each replicate to give four sub-plots each containing twenty plants, one sub-plot to be destructively harvested at each nominated date. Details of these techniques for brassicas were described and discussed by Chowings (1974); for carrots by Bedford (1975); and for celery, lettuce, leeks, onions and sweetcorn by Chowings (1975).

The points relating to good field experimentation should be observed; these include use of guard rows, uniform treatment of site and uniformity of cultural operations. The trials should be planned carefully in order to gain the maximum amount of information because information not recorded at the time of the trial cannot be retrieved later.

Husbandary techniques used in cultivar trials should normally be in accordance with local commercial practice. Where more than one centre is being used for a trial it is important that each adheres to local customs, and dates of cultural operations cannot be dictated because weather and soil may well differ between centres. But if irrigation, herbicides, fertilizers and other inputs are being incorporated, trial co-ordinators should ensure that stages of crop, moisture deficits and rates of application are clearly defined and adhered to as far as is practical.

CULTIVAR CLASSIFICATION

The cultivar trials discussed earlier in this chapter emphasize the usefulness of cultivars for particular purposes. During the course of cultivar trials, and also as a result of observations made during testing for DUS, a great deal of information is obtained about individual seed stocks. This information can be used to group cultivars according to common features and to classify them so that each cultivar can be clearly described and identified. During the course of this work the existence of synonyms and near synonyms can be clarified. In addition to synonyms it is possible that homonyms are found (homonyms are different cultivars which have the same name).

The morphological and other characters which assist in the identification of individual cultivars are of particular interest to the seed producer. The knowledge of a cultivar's observable distinguishing characters enables workers to inspect plants for trueness of type before they are used for seed production. But because some characters (for example flower colour) are only visible at certain stages of the plant's development it is useful to have a record (or cultivar description) which assists in the identification at several different stages. In this way the presence or absence of specific visible characters at specific stages will usually provide clear evidence of the cultivar's identification.

These visual characters, such a leaf shape, hypocotyl colour and flower colour, used for identification are referred to as *discontinuous* characters and may be readily determined at the appropriate plant stage. Characters in this category should be easily identified without disturbing the plant and as far as possible they should not be significantly influenced by the environment. Table 1.4 shows the use of discontinuous characters in the classification of runner bean cultivars (*Phaselous coccineus*) (Bowring 1970).

There are many occasions when attempts are made to classify vegetable cultivars where the classification cannot be based on discontinuous characters alone. Other characters, such as shape and length of pod, size of fruit and colour intensity of a specific part of the plant have to be used. These are examples of continuous characters and are so called because they are all present in the plants under scrutiny, but it is the degree or extent to which the character is present that is used for classifications. Many discontinuous characters in cultivars of the same species can be observed and categorized arbitrarily, and the extent to which this succeeds may sometimes depend on the experience and training of the observers. However, when the differences are relatively small, measurements of the expression of the character have to be taken on a number of individual plants and statistically analysed.

An example of the information from the statistical analysis of continuous characters using certain pod and seed measurements from the runner bean cultivars listed in group A of Table 1.4 is shown in Table 1.5.

Important factors to consider when deciding on morphological characters to use in cultivar classification of vegetables include the stability of characters from one generation to the next; unstable characters are not useful. Another consideration is the extent to which the environment used for testing can be the same each year and at each centre where the plants are grown. Major

SEED CERTIFICATION

When a vegetable grower has decided on the most appropriate cultivar to use for a specific purpose it is desirable that he be supplied with genuine seed or plant material of that cultivar. A range of schemes which certify the authenticity of the seed sold have evolved in different parts of the world, all of which provide some form of assurance that the seed supplied is actually of the cultivar that it is claimed to be. This ensures verification of the seed stock without the farmer or grower having to wait to see the crop which is grown from it.

These certification schemes are generally organized on a national basis and a wide range of schemes providing verification of sowing and planting materials have been operating within some countries since the 1920s. It was not until the 1950s that there were steps towards co-ordinated efforts between countries. These were instigated by the Organization for Economic Co-operation and Development (OECD) and have become the basis on which most seed certification schemes have been modelled (OECD 1977).

The OECD scheme is not confined to countries within Europe; any country which is a member of the United Nations or its specialized agencies, such as FAO, can participate. However, countries which do agree to participate in the OECD Vegetable Seed Scheme are obliged to abide by the rules.

Individual schemes within countries are usually organized on a national basis. Their primary aim is to check the crop from which the seed is produced and link this verification with agreed minimum standards of other important features of the seed-lot. These include health, potential germination and mechanical purity. Thus the structure of seed certification schemes is complex. This is necessary in order to ensure that all the facets of seed quality are included in the framework of the scheme. Despite the apparent complexity, each part of a certification scheme can be clearly defined or described.

Components of a seed certification scheme

The different requirements and standards which are examined separately but which contribute to the composite assessment of a seed-lot for certification purposes vary from country to country but generally the following components are included:

1. A *designated authority*, which is appointed by the national government, is responsible for implementing the rules of the scheme on behalf of the government. It acts as the inspectorate of the different parts of the scheme and co-ordinates the findings of the required tests and observations.
2. Cultivars are accepted into the scheme only when they have been shown to be of significant agronomic value. This value is demonstrated in official trials.
3. The breeder, or the institute which bred the original cultivar, is responsible for the maintenance of the cultivar and supply of breeder's stock for further multiplication.
4. Each generation of seed is clearly defined, viz.:

Pre-basic This is seed material at any generation between the parental material and *basic seed*.

Basic seed This is seed which has been produced by, or under the responsibility of, the breeder and is intended for the production of *certified seed*. It is called *basic seed* because it is the basis for certified seed and its production is the last stage that the breeder would normally be expected to supervise closely.

Certified seed This is the first generation of multiplication of basic seed and is intended for the production of vegetables as distinct from a further seed generation. In some agricultural crops there may be more than one generation between basic and certified seed, in which case the number of generations of multiplication after basic seed is stated, e.g. first or second.

Standard seed This is seed which is declared by the supplier to be true to cultivar and purity, but is outside the certification scheme.

5. The agronomic requirements to be observed in planning and production of the seed are defined for specific crops. These include points such as administrative checks on the history of the site, its distance from other specified crops and the number of years since previous crops of the same or related species were grown on the same site.

6. (a) Laboratory and control plot tests of the stock seed used for production of certified seed production.

    (b) Inspection of the seed crop production field and observations to be made by official inspectors employed by the designated authority. These include trueness to type, isolation and freedom from specific weeds and seed-borne diseases.

    (c) Sampling techniques to be employed.

    (d) Tests to be done in the laboratory and on field plots with samples to check the identity and purity of the cultivar.

    (e) Tests to be done in the laboratory to determine germination, purity and presence of specific seed-borne pathogens.

The success of a seed certification scheme depends on the collection and compilation of evidence from several aspects of seed production and quality control. Furthermore it also relies on demands for certified seed by vegetable growers. Thus a successful certification scheme depends on the ability of a country to produce seed of high quality in sufficient quantities to meet market requirements. It is the culmination of successful seed programmes which over several years have established the required infrastructure of a seed industry. If seed certification schemes in a country aim higher than the industry's present genuine ability, then not only will they fail but their shortcomings or failure will be detrimental to any future development of seed legislation or seed control in that country.

REFERENCES

Bedford, L. V (1975) A comparison of single row and larger plot techniques for variety performance trials of carrots, *J. Nat. Inst. Agric. Bot.* **13**, 349–54.

Bowring, J. D. C. (1970) The identification of varieties of runner bean (*Phaseolus coccineus* L.), *J. Nat. Inst. Agric. Bot.* **12**, 46–56.

Chowings, J. W. (1974) Vegetable variety performance trials technique – brassica crops, *J. Nat. Inst. Agric. Bot.* **13**, 168–85.

Chowings, J. W. (1975) Vegetable variety performance trials technique – celery, lettuce, leeks, onions and sweet corn. *J. Nat. Inst. Agric. Bot.* **13**, 355–66.

Ellis, J. R. S. and Bedminster, C. H. (1977) The identification of UK wheat varieties by starch gel electrophoresis of gliadin proteins, *J. Nat. Inst. Agric. Bot.* **14**, 221–31.

FAO (1981) *FAO Seed Review 1979–90.* FAO, Rome.

Feenstra, W. J. (1960) Biochemical aspects of seed-coat inheritance in *P. vulgaris,* Meded. Landbhoogesch, Wageningen **60**(2), 1–53.

Feistritzer, W. P. (1981) The FAO Seed Improvement and Development Programme (SIDP), *Seed Sc. Technol.* **9**(1), 37–45.

Feistritzer, W. P. and Kelly, A. F. (eds) (1978) *Improved Seed Production.* FAO, Rome.

Fennell, J. F. M. and Dowker, B. D (1979) Screening cultivars for characteristics required in breeding Autumn-sown onions, *J. Nat. Inst. Agric. Bot.* **15**, 104–12.

Gilmour, J. S. L. (1969) (International Code of Nomenclature of Cultivated Plants – 1969. The International Bureau for Plant Taxonomy and Nomenclature, Utrecht.

Grubben, G. J. H. (1977) *In Tropical Vegetables and Their Genetic Resources,* H. D. Tindall and J. T. Williams (eds). International Board of Plant Genetic Resources, Rome.

Hegnauer, R. (1967) Chemical characters in plant taxonomy: some possibilities and limitations, *Pure Appl. Chem.* **14**, 173–87.

Holland, H. (1957) Classification and performance of red beet, *Ann. Rep. Nat. Veg. Res. Sta.* 16–42.

Hollingsworth, D. F. (1981) The place of potatoes and other vegetables in the diet. In *Vegetable Productivity,* C. R. W. Spedding (ed.). Macmillan, London.

Johnson, A. G. (1956) Spring cabbage varieties, *Ann. Rep. Nat. Veg. Res. Sta.* 17–34.

Leenders, H. H. (1967) The function of the international seed trade federation (FIS) in the international seed trade, *Proc. Int. Seed Test. Ass.* **32** (2), 245–57.

McKee, G. W. (1973) Chemical and biochemical techniques for varietal identification, *Seed Sci. Technol.* **1**, 181–99.

Nieuwhof, M. (1969) *Cole Crops.* Leonard Hill, London.

OECD, (1977) *OECD Scheme for the Control of Vegetable Seed Moving in International Trade.* Organization for Economic Co-operation and Development, Paris.

Oomen, H. A. P. C. and Grubben, G. J. H. (1978) *Tropical Leaf Vegetables in Human Nutrition.* Royal Tropical Institute, Amsterdam and Orphan Publishing Company, Willemstad, Curacao.

Rowlands, D. G. and Corner, J. J. (1963) Biochemical differences as varietal characteristics for peas, beans and spinach, *Hort. Res.* **3**, 1–11.

Silvey, V. (1978) Methods of analysing NIAB variety trial data over many sites and several seasons, *J. Nat. Inst. Agric. Bot.* **14**, 385–400.

Singh, K., and Thompson, B. D. (1961) The possibility of identification of vegetable varieties by paper chromatography of flavonoid compounds, *Proc. Amer. Soc. Hort. Sci.* **77**, 520–7.

UPOV (1973) International Union for the Protection of New Plant Varieties. *General Introduction to the Guidelines for the Examination of Distinctness, Homogeneity and Stability of New Varieties of Plants.* UPOV, Geneva.

UPOV (1974) *International Convention for the Protection of New Varieties of Plants.* UPOV, Geneva.

Watts, L. E. (1955) Synonymy in lettuce varieties, *Ann. Rep. Nat. Veg. Res. Sta.* 16–36.

Watts, L. E. (1980) *Flower and Vegetable Plant Breeding.* Grower books, London.

Watts, L. E. and George, R. A. T. (1958) Classification and performance of varieties of radish, *Ann. Rep. Nat. Veg. Res. Sta.* 15–27.

Webster, T. (1974) Vegetable varieties – Why distinct, uniform and stable? *J. Nat. Inst. Agric. Bot.* **13**, 160–7.

FURTHER READING

Copeland, L. O. (1976) *Principles of Seed Science and Technology,* Chapter 13, pp. 297–318. Burgess Publishing Company, Minneapolis.

Delouche, J. C. and Potts, H. C. (1971) *Seed Programme Development.* Mississippi State University.

Douglas, J. E. (1980) *Successful Seed Programmes.* Westview Press Inc., Boulder, Colorado.

George, R. A. T. (1978) The problems of seed production in developing countries, *Acta Horticulturae* **83**, 23–9.

George, R. A. T. and Evans, D. R. (1981) A classification of winter radish cultivars, *Euphytica* **30**, 483–92.

Hawkes, J. G. (ed.) (1968) *Chemotaxonomy and Serotaxonomy.* Academic Press, New York.

Horne, F. R. and Kelly, A. F. (1967) The OECD certification schemes, *Proc. Int. Seed Test. Ass.* **32** (2), 259–73.

Innes, N. L. (1983) *Breeding field vegetables.* Asian Vegetable Research and Development Center, 10th Anniversary Monograph Series, Shanhua, Taiwan, Republic of China. 34 pp.

ISTA (1967) *Proceedings of International Seed Testing Association,* Seed Legislation Number, **32** (2).

Kelly, A. F. (1982) Seed certification. In *Proceedings FAO Seed Symposium, Nairobi,* pp. 405–12. FAO, Rome.

Sneddon, J. L. (1969) Identification features of white seeded stringless varieties of French beans (*Phaseolus vulgaris* L.), *J. Nat. Inst. Agric. Bot.,* **11**, 476–98.

Spedding, C. R. W. (ed.) (1981) *Vegetable Productivity. The Role of Vegetables in Feeding People and Livestock.* Symposia of the Institute of Biology No. 25, Macmillan, London.

Thomson, J. R. (1979) *An Introduction to Seed Technology,* Chapter 12, pp. 157–67. Leonard Hill, London.

Tindall, H. D. (1968) *Commercial Vegetable Growing.* Oxford University Press, Oxford.

Walker, J. T. (1980) Philosophies affecting the spread and development of seed production in the world. In *Seed Production,* P. D. Hebblethwaite (ed.), pp. 15–34. Butterworths, London, Boston.

2 PRINCIPLES OF SEED PRODUCTION

Before seeds can be produced from vegetables it is necessary for the crop to flower. Plant physiologists have made detailed studies of the requirement for flowering in several plant species and the reader is referred to Bleasdale (1973) and Wareing and Phillips (1981) for fuller details.

Some plant species pass from the vegetative phase to the reproductive phase with no special requirement or stimulus whereas in others there is a clearly defined transition between the two phases. The initial phase before the plant is receptive to the external flowering stimulus is referred to as the juvenile phase. Plant physiologists often refer to the conclusion of the juvenile phase as puberty. The attainment of the required physiological stage or 'age' is related to factors such as stage of growth, e.g. the number of leaves rather than actual age as described, for example, by the number of days from sowing.

Species which have a special physiological requirement to pass from the vegetative stage to puberty are generally either dependent on daylength (photoperiod) or have a low temperature requirement (vernalization).

PHOTOPERIODISM

Plants can be classified into three main groups according to the specific durations of light and dark requirement in each twenty-four hour period or cycle in order to initiate flowers, viz. short-day plants, long-day plants and day-neutral plants.

Short-day plants

This group includes species which will not flower unless the light period is shorter than a particular critical time. Many of the species originating from the low latitudes either side of the Equator where the natural daylength does not exceed fourteen hours are short-day plants, for example African spinach (*Amaranthus* spp.).

Long-day plants

These include plants which will flower only when the light period is greater than a critical time. Examples include radish (*Raphanus sativus* L.) and European spinach (*Spinacea oleracea* L.).

Day-neutral plants

This group, which are sometimes referred to as indeterminate, do not have a specific daylength requirement. However, they usually need to reach a specific stage of growth before flowers are initiated; examples include tomato (*Lycopersicon esculentum* Mill.) and broad bean (*Vicia faba* L.).

VERNALIZATION

Some vegetable species do not initiate flowers until the plant has received a cold stimulus. Examples include celery (*Apium graveolens* L. var. *dulce* Mill.) and some of the cole crops or brassicas (*Brassica oleracea* L.). These species are biennials and flower only once in the spring or early summer following a cold stimulus.

In some species the imbibed or germinating seeds can receive sufficient cold stimulus to promote flowering, and cultivars of a species even differ in their response; such is the case with turnip (*Brassica rapa* L.) and white mustard (*Sinapis alba* L.). In other species the plants have to achieve a certain size or stage of development before responding to low temperatures. This is taken into account when deciding on the sowing time of biennials for seed production to ensure that the plants will respond to the low winter temperatures. For example, onion plants for the 'seed to seed' method have to reach a larger size by the start of winter than an autumn sown crop overwintered to produce a bulb crop. If the latter crop were sown too early and hence overwintered as relatively large plants there would be a high percentage of bolters the following summer.

Interaction between vernalization and daylength

There are some vegetable species which require vernalization and also long days following vernalization before the inflorescences emerge. Examples are beetroot (*Beta vulgaris* L.) which requires a daylength of at least twelve hours following vernalization and cabbage (*Brassica oleracea capitata* L.) which shows a gradation of time to flowering after vernalization in relation to the amount of cold treatment. Heide (1970) studied this phenomenon in cabbage and concluded that not only does the response to chilling increase with the plant's age but that at the optimum temperature of 5 °C only 3–4 weeks' vernaliz-

ation is required compared with 24 weeks at 12 °C. Earlier work by Ito and
Saito (1961) established that different cultivars of cabbage have different low
temperature requirements. Thomas (1980), investigating the flowering response of seven cultivars of Brussels sprouts, concluded that although each
cultivar needed approximately thirty nodes (i.e. leaves plus leaf initials) before
reaching puberty, the quantitative response to low temperature was different
for early-, mid- and late-maturing cultivars.

Other responses influenced by daylength

Daylength can affect vegetative processes or developments in plants in addition to controlling flower initiation. For example, the formation of tubers
in the Jerusalem artichoke (*Helianthus tuberosus* L.) and bulb formation in
onion are controlled by long days. The daylength response of onion cultivars
is important in choosing those appropriate for bulbing in different areas of
the world and it follows that for seed production, cultivars must be grown
where adequate bulbing occurs in order to verify this character.

Magruder and Allard (1937) first described the effect daylength had on bulb
formation of some onion cultivars; more recently Austin (1972) has shown
that different cultivars' bulbing responses to daylength remained in the same
relative order over a wide range of photoperiods.

THE USE OF GROWTH REGULATORS TO PROMOTE FLOWERING

Plant physiologists have investigated the use of growth regulators to promote
flowering in a large number of plant species. So far gibberellic acid has been
the most promising material to be investigated and seems especially useful
in those biennials which form a rosette of leaves in their first season and have
a vernalization requirement to bolt in their second year (Luckwill 1981). Several workers have investigated the use of growth regulators to replace or reduce the appropriate cold stimulus of a range of species. According to
Wareing and Phillips (1981) the vegetable species with a chilling requirement
which flower under non-inductive conditions in response to gibberellic acid
include celery (*Apium graveoleus* L. var. *dulce* Mill.) cabbage (*Brassica oleracea
capitata* L.) and carrot (*Daucus carota* L. subsp. *sativus* (Hoffm.) Thell.).

In the tropics there has been an interest in the production of seeds from
vegetable species which normally have a vernalization requirement. Kahangi
and Waithaka (1981) have reported that GA_3 applied to some brassica cultivars grown at an altitude of 1941 m at Kabete, Kenya, replaced the vernalization requirement. While the development of this technique can be a useful
aid to seed production of some species in the tropics, it is important that the
full range of morphological characters of a vegetable crop cultivated for seed
production should be confirmed before bolting is induced. This is especially
important in the production of basic seeds.

SEX EXPRESSION AND THE USE OF HORMONES

Techniques using growth hormones have been developed to influence the sex expression of flowers in some vegetable species. This has become especially important in the production of F_1 hybrids in several seed crops and also in the production of seeds from all female lines of some cucurbits, especially cucumbers. The physiological background to this is discussed by Wareing and Phillips (1981).

In some cultivars of cucumber (*Cucumis sativus* L.) plant breeders have developed 'all female' lines. While this has several advantages for the cucumber producer, including the production of seedless fruits (as a result of parthenocarpy), the seed producer has the problem of producing staminate (male) flowers. Conversely, the production of any staminate flowers has to be suppressed on the lines used as the female parent during hybrid seed production. The practical use of growth regulators in these circumstances is discussed in Chapter 10 in the sections on cucumbers and *Cucurbita* spp. The different methods used to achieve male sterility are given below in the sections relating to F_1 hybrids and the use of gametocides.

POLLINATION AND FERTILIZATION

The morphology of different types of flowers, pollen formation and the process of fertilization have been documented by several authors and the reader is referred to Copeland (1976) or North (1979).Detailed flower morphology of the important vegetable crops in the world is given by McGregor (1976). The theory of incompatibility systems and their practical implications are discussed by Lewis (1979).

The transfer of pollen to the stigma is achieved in the flowering plants by either wind, animals or water. In practice, as far as vegetable seed production is concerned, it is only the transfer by wind or animals which normally concerns us, and the most important pollinating agents in the animal kingdom are the insects. Flowers which are wind pollinated are said to be anemophilous, e.g. sweet corn (*Zea mays* L.) and European spinach (*Spinacia oleracea* L.). Flowers which are insect pollinated are entomophilous and examples include most of the cole crops (*Brassica oleracea* L.), carrot (*Daucus carota* L.) and onion (*Allium cepa* L.). Some of the species which may on occasions be cross-pollinated by insects are in practice largely, if not completely, self-pollinated. This occurs as a result of flower morphology which is highly adapted to self-pollination, and examples include the garden pea (*Pisum sativum* L.), dwarf bean (*Phaseolus vulgaris* L.), lettuce (*Lactuca sativa* L.) and tomato (*Lycopersicon esculentum* Mill.). The method of pollination for each seed crop is given in Chapters 7–16.

The use of insects to increase seed yield

A very large number of insect species are directly involved in pollination, for example Bohart and Nye (1960) identified 334 insect species visiting carrot flowers and Hawthorn *et al.* (1960) identified 267 insect species on onion flowers.

The two most important insect orders concerned with the pollination of vegetable seed crops are Hymenoptera which includes the ants, bees and wasps, and Diptera, the large order which includes the flies (Fig. 2.2). The level of pollinating insect activity in an anemophilous species grown for a seed crop will have a direct effect on seed yield. In many instances the seed produced relies entirely on natural insect pollinations in addition to the roving honeybees maintained by bee keepers. Bohart *et al.* (1970) have reported that in the USA large areas of alfalfa seed production in Idaho and Oregon have 'drained off' populations of pollinating insects. On several occasions in the Sudan the author has observed virtually no insect activity on fields of onion in full flower; this paucity of insects is attributed to the very frequent applications of insecticides to the cotton crops in the area. Many specialized vegetable seed producers, especially in the USA, ensure adequate pollinator activity while their crops are in flower by arranging for colonies of honeybees to be supplied on a contract basis (Fig. 2.1).

Pollinating insect activity can be increased by improving the micro-climate. Drifts of wind-pollinated crops such as maize or sweet corn are planted by some seed producers to improve the environment. Crops such as sunflowers

Fig. 2.1 Colonies of honeybees supplied by contractors to seed producers in the USA.

Fig. 2.2   A blowfly visiting a carrot inflorescence

may also be used, especially if their flowering does not coincide with the vegetable seed crop.

A wide range of pesticides is used in modern farming systems and many of them are toxic to bees and other beneficial pollinating insects. The protection of honeybees from pesticides has been reviewed by Atkins *et al.* (1977). Generally, co-operation between the beekeepers and the pest control industry will ensure that only the safest recommended material is used for a specific purpose. Supplementary hives should be located on the perimeters of flowering crops and placing them in position two to three days after pesticide applications will minimize the danger. Pesticides toxic to bees should not be applied while bees or other important pollinators are foraging, and drift from spraying operations on to colonies or flowering crops should be avoided.

Other factors which can adversely affect the level of pollination include rainfall. Ogawa (1961) reported that continuous rainfall during onion flowering reduced the seed set significantly and that the reduction of seed yield was proportional to the duration of continuous rain. Dry periods of up to six hours improved the seed set. Local weather not only affects the pollinating insect activity but may be responsible for a reduced rate of stamen dehiscence.

Pollen beetles

The pollen beetles (*Meligethes* spp.) are largely confined to attacking the cole crops and related species. The small blue-black beetle overwinters in hedgerows, but in warmer weather migrates to crucifers. This pest is often not noticed until 'blind' stalks are observed, by which time it is too late for an effective control. Spraying with appropriate insecticides should be done when the plants are in the early bud stage (Anon 1981).

Pollination in confined spaces

There are occasions when plants for seeding are cultivated in, or transferred to, enclosed structures such as cages, polythene tunnels or greenhouses. These structures are discussed in greater detail in Chapter 3. Although the isolation provided by the structure normally eliminates the possibility of insects transporting undesirable pollen from outside, it is necessary to ensure that adequate pollination occurs within the enclosure. In some instances hives or smaller honeybee colonies are introduced and the pollinating insects work effectively. However, in some instances, such as relatively small cages, bees have a strong tendency to attempt to escape from the enclosure, and die in the process. Bohart *et al.* (1970) reported that honeybees operating in cages depressed the potential seed yield of onions. They found evidence of physical injury to the stigmatic surfaces and that the bees stripped pollen from the stigmas. Faulkner (1971) reported that bees tended selectively to self- or sib-pollinate each of the two inbred parents in Brussels sprout hybrid seed production, in contrast to blowflies which pollinated at random and assured a higher degree of cross-pollination between the parent lines. The efficiency of blowflies as pollinators of brassica crops was reported by Faulkner (1962) and they are used as pollinating agents for some seed crops of Cruciferae, Umbelliferae and Alliaceae when grown in enclosed environments (Fig. 2.2). Smith and Jackson (1976) described the methods of providing adult blowflies for pollination work. So far no work has been reported on the deliberate encouragement or production of blowflies for pollination in the field.

Effect of environment on pollination method

It is possible for a change in emphasis of a species' pollination method to occur when it is grown outside its natural habitat. This can result in a degree of instability which is not always appreciated. For example, the tomato flower is normally accepted as being completely, or at least predominantly, self-pollinated, but work done in California by Rick (1950) has demonstrated that there can be a significant amount of cross-pollination in some locations when specific pollen vector insects are present. *Pisum sativum* L. (the garden pea) is normally accepted as being completely self-pollinated, but Harland (1948) showed that when grown in Peru (an area outside its normal habitat) there was up to 3.3 per cent cross-pollination.

The above examples demonstrate that the predominant pollination method may modify with provenance of seed production.

F₁ HYBRIDS

A F_1 hybrid is produced by crossing two distinct lines. In practice the two parent lines are the result of inbreeding. As far as commercial seed production is concerned F_1 hybrid cultivars are the result of crossing two inbred lines which have been maintained under the control of plant breeders and which are known to produce a desirable hybrid.

There has been an increasing number of F_1 hybrid cultivars available on the market in the last two decades since F_1 hybrid cultivars of vegetables were pioneered by Japanese plant breeders in the 1930s (Yamashita 1973).

The exploitation of hybrid vigour in vegetables, with special reference to F_1 hybrids developed for India, has been reviewed by Choudhury (1966). There are in different areas of the world seed companies specializing in the production and marketing of F_1 hybrid cultivars of vegetables.

The advantages of F_1 hybrid cultivars include uniformity, increased vigour, earliness, higher yield and resistance to specific pests and pathogens, although these factors are not always all present in any one cultivar. Crops which have a noticeably high number of F_1 cultivars widely adopted by vegetable producers include Brussels sprouts, cabbage, cucumber, marrows, onion, squashes and tomato.

Theoretically all the plants in a F_1 hybrid cultivar resemble each other exactly, but because some self-pollination of the female parent used in the cross may take place, some plants which are not F_1 hybrids may occur and are usually morphologically different. These off-types in a F_1 hybrid are usually the result of accidental self-pollination of the female parent and are generally known as 'sibs'. These are so-called because they are the result of 'sister' or 'brother' plants crossing or selfing within the female line. The incidence of sibs in seed production of F_1 hybrid Brussels sprouts and the influence of pollinating insects on the percentage in a seed-lot have been investigated by Faulkner (1978), who demonstrated that blowflies pollinate in a more random manner than honeybees; the flies drastically reduced the percentage sibs under glass and polythene and to a lesser extent under cages.

In addition to the problems associated with sibs in F_1 hybrid seed lot, there are increased production costs compared with open-pollinated cultivars due to factors such as the development of the initial breeding programme, subsequent maintenance of the inbred parents, extra land required to allow for male parents, care with sowing, isolation and harvesting, high labour input when flowers of the female parent have to be emasculated (especially when hand emasculation is necessary) and the lower yield sometimes experienced with production of F_1 hybrid cultivars.

The different systems utilized in the production of F_1 hybrids have been listed by Wills and North (1978) and are summarized below. (Details of the principles involved are described by Frankel and Galun (1977).)

Genetically controlled male sterile lines, e.g. tomato, carrot and onion.
Self-incompatibility, e.g. Brussels sprouts.
Gynoecious lines, e.g. cucumber.
Dioecy. Use one line of male and other of female, but rogue out plants showing male flower characteristics in female line, e.g. spinach (*Spinacea oleracea*).

Monoecy, male and female flowers are borne on different parts of the same plant, e.g. sweet corn.
Manually controlled, e.g. emasculation of tomato flower on female lines; 'detasselling' of male inflorescences e.g. sweet corn; the female inflorescences, 'silks', are lower on the plant.
Chemically controlled, use of gametocide (e.g. use of growth regulator to suppress male flowers of female line, as in cucurbits).

Wills and North (1978) also outlined plant breeders' proposals which suggested that the increased use of marker genes, for example green versus white perianth and glossy versus non-glossy foliage in onions, would facilitate roguing of contaminants in inbred lines.

Philosophy of production and use of F_1 hybrid cultivars

As a result of the reasons outlined above, hybrid seed of a given crop is generally more expensive than non-hybrid seed and this is generally attributed to the costs of development and production. One main advantage of a F_1 hybrid to the seed company developing and marketing it is the relative difficulty with which competitors can reproduce the cultivar. But this advantage is generally considered less important now in areas where there is an effective implementation of plant breeders' rights. It is also suggested that F_1 hybrids provide seed companies with a hold on the market because once growers have found a hybrid cultivar to be acceptable they will continue using it in subsequent years.

It is stressed that hybrid cultivars of vegetables should be judged on their agronomic and economic advantages to the grower rather than as a way by which seed companies increase their revenue.

DOUBLE CROSS HYBRIDS

The double cross hybrids are produced by crossing two pairs of inbred lines. The two resulting F_1s are then crossed to produce a double hybrid. This technique has been used in the production of some brassica and sweet corn cultivars and is better than single cross hybrids because the seed yield is generally higher, although the maintenance of the appropriate lines and crosses increases the isolation requirements.

TRIPLE CROSS HYBRIDS

This type of hybrid was described by Thompson (1964) as a method of overcoming the relatively high costs of maintaining inbred lines. Although not of major significance in vegetable crops, it has become important in the production of some of the fodder brassicas.

SYNTHETIC HYBRIDS

A synthetic hybrid is produced by the mass pollination of several inbred lines selected for their satisfactory combination ability. Random cross-pollination takes place between the different inbred lines, resulting in a mixture of hybrids. The individual inbred lines to be used for the production of these synthetic hybrids are determined by the plant breeder. Because of the random crossing which takes place there may be some variation from one season to another when the same parents are used. Cross-pollination is normally assured because of the relatively high levels of self-incompatibility of individual inbred lines. This system is used for the production of some cultivars of summer cabbage and other brassicas.

The use of gametocides to produce male sterility

The availability of genetical and cytoplasmic male sterility has been of great assistance in the development of F_1 hybrid vegetable cultivars. The mechanical or hand emasculation of flowers on the female parents is relatively time consuming and therefore costly; in addition there always exists the possiblility of human error in that all anthers are not removed. An alternative to these methods of obtaining male sterility is the use of a gametocide. Saimbhi and Brar (1978) reviewed the practical use of gametocides on vegetable crops and reported that out of fifteen chemicals tested maleic hydrazide at 100 to 500 ppm was the most effective in inducing a relatively high level of pollen sterility in eggplant, okra, sweet pepper and tomato without a detrimental influence on female fertility. The most effective concentrations of maleic hydrazide applied as aqueous solutions to foliage before anthesis reported for specific crops were 100 ppm to eggplant and onion, 100–500 ppm to tomato and 400–500 ppm to okra and sweet peppers.

Other materials which have been reported include sodium 2, 3-dichloro-isobutyrate (Mendok), but while causing a high proportion of pollen to be sterile it also produced adverse effects on plant vigour and fruit set.

Synchronization of flowering

Where the pollen donor (male) and pollen receptor (female) parts are on separate plants as in the production of hybrid seeds, it is essential that the shedding of pollen and receptiveness of the stigma occur simultaneously. Seed producers refer to this synchronization of flowering as 'nicking'. The matching of parent lines for hybrid seed production therefore depends upon their ability to 'nick', as well as on agronomic characteristics of the progeny. Pairs of inbreds which 'nick' in one location do not necessarily flower at the same time or duration in another. 'Nicking' may also vary from one season to another. Faulkner *et al.* (1977) reported that when parent lines of Brussels sprouts were tested for matching ability based on plant height, flower colour and time of flowering the cross-pollination by bees was good.

Synchronization of flowering between parent lines of some other vegetable seed crops can sometimes be assured by time of sowing of the pollen parent in relation to the female as well as successive sowings to ensure a satisfactory overlap of anthesis. Information on these requirements may be provided by the plant breeder supplying the inbred parents, but will also be learned from experience in a specific location.

IMPORTANCE OF WATER DURING ANTHESIS AND SEED DEVELOPMENT

Most experimental investigations into the water requirements of crops have been directed at the market yield of the vegetative stage. Very little work has been done into the effects of soil moisture deficit on seed yield and quality. However, there are indications from the literature which can be extrapolated by the vegetable seed producer.

A comprehensive review of crop responses to water at different stages of growth was made by Salter and Goode (1967) who examined the available information on different crop groups, including legumes, annual fruits, leafy and biennial crops. Generally the papers reviewed indicated that most annual crops have moisture-sensitive stages from flower initiation, during anthesis, and in some species during fruit and seed development. The legumes are the vegetable group which has received the most attention as to the effects of soil moisture shortages on seed yield. The literature indicates that provided there is sufficient water available prior to flowering for plant growth to proceed without permanent wilting, there is little influence on seed yield. But the plants are very sensitive to moisture stress during anthesis when adequate soil moisture, supplemented where necessary by irrigation, will generally provide maximum yield increases. Salter and Goode found that some investigators working with legumes reported an increase in the number of seeds per pod as a result of additional irrigation applied early in anthesis and that further irrigation during pod growth increased the 1000 grain weight. This general effect in the leguminous vegetables is believed to be due to a reduction in root activity when the plants are flowering. The evidence for this on annual leafy vegetables is not clear, especially as most experimental work has been directed at increasing leaf or bud yield rather than seed. The fresh market crop yield of leafy vegetables such as lettuce and cabbage is generally in proportion to the amount of water received. Some workers considered that this response of vegetative growth is a precursor to flower formation, but there is little experimental evidence to demonstrate this.

The biennial vegetable crops comprise a wide range of morphological types at the time of marketing, including root, bulb, leaf, petiole or reproductive tissue (e.g. the 'curd' of cauliflower). These different crop types usually display a response to irrigation during growth and development in the first year by an increase in size of their storage organs. There is also evidence that seed yields of these biennials increase with the available water during anthesis.

The work on the 'fruit' vegetables (e.g. Solanaceae) indicates that they are responsive to adequate available soil water once fruit setting commences. Salter (1958) reported that water shortage during anthesis and fruiting of tomato reduced fruit set and size. He found that this pattern was generally true for satisfactory vegetative development for all the indeterminate fruiting vegetables (Salter and Goode 1967).

NUTRITION AND FLOWERING

There is a tendency for there to be antagonism between the mineral nutrient requirements for optimum vegetative growth and optimum reproduction. Wareing and Phillips (1981) reported that low levels of nitrogen tend to result in earlier flowering of some long-day plants. Raut (Pers. Comm), investigating the effects of cauliflower mineral nutrition on seed yield and quality, found that when nitrogen was applied at 50, 150, 250 and 350 kg/ha, the highest level delayed flowering by up to ten days. Liaw (1982), in similar work but with dwarf beans (*Phaseolus vulgaris* L.), found that the higher levels of phosphorus increased flower number.

Fruit and seed development

The seed develops as a result of fertilization following successful pollination. Many botany textbooks give an account of fertilization with subsequent seed and fruit development. The reader is referred to North (1979) or Copeland (1976) for a plant breeder's account of this process.

REFERENCES

Anon. (1981) *Guide to the Arrangements in North Essex for the Prevention of Injurious Cross-Pollination of Seed Crops of Allium (Onion). Beta (Beet) and Brassicas.* Ministry of Agriculture, Fisheries and Food, London.

Atkins, E. L., Anderson, L. D., Kellum, D. and Neuman, K. W. (1977) *Protecting Honey Bees from Pesticides.* Leaflet 2883, Division of Agricultural Sciences, University of California.

Austin, R. B. (1972) Bulb formation in onions as affected by photoperiod and spectral quality of light, *J. Hort. Sci.* **47**, 493–504.

Bleasdale, J. K. A. (1973) *Plant Physiology in Relation to Horticulture.* Macmillan, London.

Bohart, G. E. and Nye, W. P. (1960) Insect pollinators of carrots in Utah, *Utah Agric. Exp. Sta. Bull.* **419**, 1–16.

Bohart, G. E., Nye, W. P. and Hawthorn, L. R. (1970) Onion pollination as affected by different levels of pollinator activity, *Utah State Univ. Bull.* **482**.

Choudhury, B. (1966) Exploiting hybrid vigour in vegetables, *Indian Hort.* **10**,56–8.

Copeland, L. O. (1976) *Principles of Seed Science and Technology.* Burgess, Minneapolis.

Faulkner, G. J. (1962) Blowflies as pollinators of brassica crops, *Commercial Grower* **3457**, 807–9.

Faulkner, G. J. (1971) The behaviour of honeybees (*Apis mellifera*) on flowering Brussels sprout inbreds in the production of F₁ hybrid seed, *Hort. Res.* **11**, 60–2.

Faulkner, G. J. (1978) Seed production of F₁ hybrid Brussels sprouts, *Acta Horticulturae* **83**, 37–42.

Faulkner, G. J., Smith. B. M. and Draycott. A. (1977) Matching inbred lines of Brussels sprouts for flowering characteristics as an aid to improving F₁ hybrid seed production, *Ann. Appl. Biol.* **86**, 423–8.

Frankel, R. and Galun, E. (1977) *Pollination Mechanisms, Reproduction and Plant Breeding.* Springer-Verlag Berlin, Heidelberg and New York, 281 pp.

Harland, S. C. (1948) Inheritance of immunity to mildew in Peruvian forms of *Pisum sativum, Heredity* **2**, 263–9.

Hawthorn, L. R., Bohart. G. E., Toole, E. H., Nye, W. P. and Levin, M. D. (1960) Carrot seed production as affected by insect pollination, *Utah Agric. Exp. Sta. Bull.* **422**.

Heide, O. M. (1970) Seed-stalk formation and flowering in cabbage 1. Day length, temperature and time relationship, *Meld. Norg. Landbrhisk* **49** (27), 21.

Ito, H. and Saito, T. (1961) Time and temperature factors for the flower formation in cabbage, *Tohokli J. Agri. Res.* **12**, 297–316.

Kahangi, E. M. and Waithaka, K. (1981) Flowering of cabbage and kale in Kenya as influenced by altitude and GA application, *J. Hort. Sci.* **56**(3), 185–8.

Lewis, D. (1979) *Sexual Incompatibility in Plants.* Studies in Biology No. 110. Edward Arnold, London.

Liaw, H. L. (1982) The effect of mineral nutrition on seed yield and quality in *Phaseolus vulgaris. MSc Thesis*, University of Bath.

Luckwill, L. C. (1981) *Growth Regulators in Crop Production.* The Institute of Biology's Studies in Biology No. 129. Edward Arnold, London.

Magruder, R. and Allard, H. A. (1937) Bulb formation in some American and European varieties of onions as affected by length of day, *J. Agric. Res.* **54**, 719–52.

McGregor, S. E. (1976) *Insect Pollination of Cultivated Crop Plants.* USDA Handbook No. 496. USDA, Washington.

North, C. (1979) *Plant Breeding and Genetics in Horticulture.* Macmillan, London.

Ogawa T (1961) Studies on seed production of onion. I. effects of rainfall and humidity on fruit setting, *J. Japan Soc. Hort. Sci.* **30**, 222–32.

Rick, C. M. (1950) Pollination relations of *Lycopersicon esculentum* in native and foreign regions, *Evolution* **4**,110–22.

Saimbhi, M. S. and Brar, J. S. (1978) A review of the practical use of gametocides on vegetable crops, *Scientia Horticulturae* **8**, 11–17.

Salter, P. J. (1958) The effects of different water-regimes on the growth of plants under glass. IV Vegetative growth and fruit development in the tomato, *J. Hort. Sci.* **33**,1–12.

Salter, P. J. and Goode, J. E. (1967) *Crop Responses to Water at Different Stages of Growth.* Commonwealth Agricultural Bureaux, Farnham Royal, England.

Smith, B. M. and Jackson, J. C. (1976) The controlled pollination of seeding vegetable crops by means of blowflies, *Hort. Res.* **16**, 53–5.

Thomas, T. H. (1980) Flowering of Brussels sprouts in response to low temperature treatment at different stages of growth. *Scientia Horticulturae* **12**,221–9.

Thompson, K. F. (1964) Triple-cross hybrid kale, *Euphytica* **13**, 173–7.

Wareing, P. F. and Phillips, I. D. J. (1981) *Growth and differentiation in Plants*, Pergamon.

Wills, A. B. and North, C. (1978) Problems of hybrid seed production, *Acta Horticulturae* **83**, 31–6.

Yamashita, T. (1973) Current utilization of F_1 hybrids for vegetable production in Japan. *Japan Journal Agricultural Research Quarterly* **7**, 195–201.

FURTHER READING

Bohart, G. E. (1952) Pollination by native insects, *Yearbook of Agriculture*, pp 107–121. USDA.

Bohart, G. E. and Koerber, T. W. (1972) Insects and seed production. In *Seed Biology* Volume III. T. T. Kozlowski, (ed.) pp. 1–53. Academic Press, New York and London.

Brewster, J. L. (1981) A comparison in the effects of artificial and natural lighting during cold treatment on flowering of Brussels sprouts, *Journal of Horticultural Science* **56**(3), 271–272.

Choudhury, B. and Babel, Y. S. (1969) Sex modification by chemicals in Bottle-Gourd *Lagenaria Siceraria* (Molina) Stand1, *Science and Culture* **35**(7), 320–322.

Dragland, S. (1972) Effect of temperature and day-length on growth, bulb formation and bolting in leeks (*Allium porrum* L.), *Medinger fra Norger landbrukshogskole* **51**(2), 1–25.

Faulkner, G. J. (1983) *Maintenance Testing and Seed Production of Vegetable Stocks.* Vegetable Research Trust, Wellesbourne, Warwick, pp. 62.

Free, J. B. (1970) *Insect Pollination of Crops.* Academic Press, London.

Hanisova, A. and Krekule, J. (1975) Treatments to shorten the development period of celery (*Apium graveolens* L.), *J. Hort. Sci.* **50**, 97–104.

Moore, J. F. (1959) Male sterility induced in tomato by sodium 2, 3-dichloroisobutyrate, *Science* **129**, 1738–1740.

Procter, M. and Yeo, P. (1973) *The Pollination of Flowers.* Collins, London.

Stokes, P. and Verkerk, K. (1951) Flower formation in Brussels sprouts, *Meded. Landbouwhogesch, Wageningen* **50**, 143–160.

Verkerk, K. (1954) The influence of low temperature on flower initiation and stem-elongation in Brussels sprouts, *Proc. Kon. Ned. Akad. Wetensch Serc.* **57**, 339–346.

3 AGRONOMY

SEED PRODUCTION AREAS

The main areas of seed production are based on climatic factors which ensure a relatively satisfactory environment for vegetable seed production. These factors include sufficient rainfall to ensure complete development and maturation of the seeds but a relatively dry summer and autumn with relatively little rain and wind to enable the seeds to ripen and the harvesting operations to be completed with minimal deterioration and crop loss. This is especially important for the dry-harvested vegetable seeds.

The lack of inclement weather during final stages of development and ripening is also important from the point of view of disease control as a low relative humidity with minimal rainfall and moderate temperatures minimize the development and dispersal of many pathogens (Gaunt and Liew 1981).

Additional advantageous climatic factors include a relatively mild winter to ensure minimal loss of overwintered plants of biennial seed crops, although for some species there must be sufficiently low temperatures during the winter to ensure satisfactory vernalization.

The soil quality has also played a part in the development of seed production areas as generally only those soils with a relatively high water holding capacity, while not being wet in winter, are most suited. Winter soil conditions are especially important when biennial plants are lifted for selection and re-planting or young plants are transplanted into their final seeding quarters. The nutrient status of soils can be modified by application of appropriate macro- and micro-nutrients but those soils with a satisfactory nutrient status and a satisfactory cation exchange capacity are most useful.

Thus, based on the above criteria of climate and soil, several classical examples of seed production areas can be cited and they include parts of North America especially the Pacific North-West, Northern Italy, Hungary and parts of Australasia. The oceanic effect in coastal areas prevents overdrying of unthreshed material in the field and can assist in reducing loss from shattering during harvesting. Excellent farming traditions, coupled with ability to apply sufficient quantities of satisfactory quality irrigation water as and when required, have also ensured that areas such as California lead the world's vegetable seed technology. There are many other highly developed vegetable seed production countries including the Netherlands, France and the UK, and other areas specialize in a specific crop group, for example, parts of East Africa have been utilized for the production of seed of some legume crops.

As the agricultural and horticultural industries develop in different countries so do the seed requirements. Hrabovszky (1982) discussed the possible effects of crop production in developing countries on seed development plans. There are already many areas where there have been substantial developments of vegetable seed production despite adverse climatic conditions; notable examples include Egypt, India, Israel and Kenya. This type of development frequently combines the climatic attributes such as high summer temperatures with high technical inputs such as irrigation. In addition these newer areas have depended on migrant technologists or technical aid. Mexico with its relatively arid climate and available labour is also emerging as a vegetable seed production area.

The climate in the British Isles is not suitable for the production of seeds of all vegetable crops on a commercial scale although the drier areas of East Anglia and notably Essex have become important for the production of some crops, such as some of the Chenopodiaceae, Cruciferae and Leguminosae.

The technological advances in protected cropping, especially the development of plastic tunnels and modern irrigation systems, are leading to further developments in seed production in temperate areas where the climate is otherwise disadvantageous. Faulkner (1983) described the use of plastic tunnels for the seed stock maintenance programme of a range of vegetable crops including some of the brassicas, celery, lettuce, onion and parsnip.

Overall, a wide range of climatic, technological and economic factors superimposed on a high standard of agronomy dictate the success of current vegetable seed production areas, and while it is always interesting to analyse the reasons for the success of existing regions it is not so easy to predict which new vegetable seed areas may emerge in response to the requirements of the twenty-first century.

THE USE OF SHELTER AND WINDBREAKS

Within each of the production areas there are local gradations of climate which result from differences in altitude, topography, proximity to the coast and presence of natural shelter belts such as woodland. Further improvement in a seed crop's micro-climate can be provided by windbreaks. These are especially useful for vegetable seed production where individual blocks of a seed crop are not always as large as some for agricultural crops.

Solid barriers such as walls divert the windflow but cause turbulence which can damage the plants but windbreaks which offer approximately 50 per cent obstruction provide a relatively extensive shelter with the minimum of wind gusts. A permeable windbreak can decrease the windspeed in a horizontal direction downwind for a distance equivalent to up to thirty times its height although the most useful shelter is within a horizontal distance on the leeward side of approximately ten times the height of the shelter (Fig. 3.1).

Shelter belts are also beneficial to the individual plants in the crop because water loss by transpiration and evaporation from the soil is reduced. There is less leaf damage by bruising, and the protective waxy layer on the leaf surfaces of species such as peas and onions remains more intact. In coastal

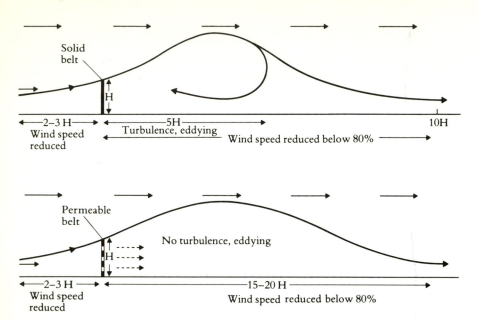

Fig. 3.1   Diagrams showing the effects of solid and permeable shelter (from (Webster and Wilson 1980)

areas windbreaks reduce the incidence of scorch from wind-borne salt originating from sea spray. There is usually a slight increase in soil temperature which is especially useful in temperate areas. Soil erosion from 'blowing' is also reduced.

In addition to the above effects the improved micro-climate will enhance flower development, increase insect activity and hence pollen transfer in entomophilous crops, and decreased mechanical damage to flowers will result in more fertilization.

Some specialist vegetable seed producers believe that the incidence of undesirable cross-pollination between different blocks of cross-compatible crops is reduced if separated by belts of maize or other suitable crops. In practice ten rows or so of these temporary shelter crops will increase the activity of bee colonies within the separate blocks and so reduce the amount of out-crossing to other blocks. This is especially useful in crops such as eggplant and peppers where up to 10 per cent out-crossing with other blocks may otherwise be expected. In these instances shelter belts are not a substitute for appropriate isolation distances (see below) but rather enhance them.

Dark (1971), investigating the cross-pollination of sugar beet, suggested that hedges may affect the incidence of wind-borne pollen in a plot. He proposed that work should be done to see if the sharp drop in pollen concentration on the windward side of a hedge and the gradual build-up again downwind from the hedge were effects of wind deflection or deposition of pollen into the hedge.

There are many different types of windbreaks, but they can be regarded as either permanent or temporary.

Permanent windbreaks

These are provided by planting lines or belts of single or mixed species of trees which are tolerant of local conditions and have a growth rate suited to quick establishment. A wide range of species is used: evergreens such as conifers and *Eucalyptus* spp. and deciduous species such as *Populus* spp., *Salix* spp. and *Tamarix* spp. Tolerance of the species to salt should be taken into account where high soil salinity is known to be a problem. In many arid areas appropriate species are planted either along water courses and irrigation channels or large land areas are subdivided into appropriate size plots by lines of trees.

Temporary shelter

This can be of two types, viz. living plants or manufactured materials. Usually with living plant materials, another crop is grown for its own value and has the dual role of providing shelter for an adjacent crop. Examples depend upon the farming system but *Zea mays* L. (either maize or sweet corn), *Helianthus annuus* L. (sunflower) or the cultivated *Sorghum* spp. are widely used. Occasionally in small-scale operations climbing plants such as *Phaseolus spp.* on a framework provide shelter.

Other forms of temporary shelter can be provided by erecting screens of manufactured materials such as plastic mesh or hessian. Plastic materials should contain ultra-violet (UV) light inhibitors to prolong their useful life by minimizing UV degradation. These types of erection are relatively low but have the advantage of immediate effect and they do not compete with crops for water and nutrients.

Possible disadvantages of living windbreaks

The roots of the species used may enter water courses and drains with the result of water loss or blockages. Dense windbreaks can reduce the available phytosynthetic light and will compete for water and nutrients.

Living windbreaks can be hosts to a wide range of pests and pathogens, for example some *Populus* spp. are alternative hosts to the lettuce root aphid (*Pemphigus bursarius* L.) and *Salix* spp. to the willow-carrot aphid (*Cavariella aegopodii* scop.) and *Salix* spp. which is the vector of carrot motley dwarf virus.

Although the relatively dense windbreaks offer shelter to a range of bird species there is no experimental evidence to suggest that they lead to an increase in crop loss from bird damage.

PREPARATION, SOWING, PLANTING AND MOTHER PLANT ENVIRONMENT

In general the principles and practices to establish the crop are the same as for the production of vegetables for market outlets. But as the final objective is to obtain seeds to be used for the production of further crop generations, it is important that every care be taken to avoid admixture of seeds or plant material at all stages and to produce a typical crop of the cultivar so that its genetical quality or 'trueness to type' can be fully evaluated.

COVER CROPS

In some areas of the world vegetable seeds are produced as cover crops in plantations. This is relatively common in the tropics and is partly a reflection of the level of seed industry development and partly because of existing agricultural crop systems, including a strong tradition to provide shade for some crops. Young plantation crops such as palm, citrus and bananas offer some shelter and in addition there is sufficient arable space for vegetables in the early life of the plantation. Some legume seed crops are grown by this system under date palms (Fig. 3.2).

CROP ROTATION

Satisfactory intervals between related or similar crops is standard agronomic practice and the main reasons for crop rotation include plant nutrition, maintenance of soil physical conditions and minimizing the risk of soil-borne pests, diseases and weeds common to individual crops or crop groups. In addition to these generally accepted reasons for crop rotation, the seed producer has to minimize the risk of plant material or dormant seeds remaining from previous crops, which are likely to cross-pollinate or make an admixture with the planned seed crop. In practice, therefore, attention must be paid to the number of years since a related crop was grown in the same soil. These periods are stringently adhered to for pre-basic, basic or certified seed. In the UK, for example, a period of five years must elapse between the production of a red beet crop and any *Beta* spp. This period is reduced to three years if the previous crop was transplanted and not direct sown. The UK regulations for the different crops entered for pre-basic, basic or certified seed are also enforced for a related crop which failed for any reason. The only possible exceptions are if satisfactory partial soil sterilization has been done or the crop is grown under glass in a substrate system which cannot be contaminated by volunteer plants or infected by specific pathogens.

Fig. 3.2 Broad bean seed production as a cover crop in a date palm plantation.

THE USE OF CAGES AND PROTECTED STRUCTURES

Cages

The primary reason for the use of cages is to provide isolation by the exclusion of pollinating insects. Having achieved this isolation it is then necessary to either hand pollinate the flowers or provide appropriate insects, unless the crop is largely self-fertilizing. The provision of insects was discussed in Chapter 2. Cages do not exclude stray pollen of wind pollinated crops.

Large cages are sometimes used by seed producers for the final quantity of plants from a positive selection for basic seed production. Cages for this purpose may be sufficiently large for up to 100 plants. Smaller cages are widely used by plant breeders. Some seed producers cover a wooden greenhouse structure with netting to maintain an ambient atmosphere but provide isolation. This type of structure is also useful for the production of virus-free seed stocks of crops, such as lettuce, provided that the mesh is aphid proof (Fig. 3.3).

Fig. 3.3   Plastic tunnels (left) and structures clad with plastic netting (right).

Greenhouses

Both glasshouses and plastic tunnels are used for seed production. The former have been used both as a convenient working area, for an improved environment and for isolation. Since the advent of large walk-in plastic tunnels they have been increasingly used for seed production. They are especially useful in temperate areas as the protected environment for both crop and pollinating insects will produce good yields of high quality seeds (Fig. 3.3). Plants are usually grown direct in the soil substrate; either sown *in situ* or material transferred from field selections can be re-planted in them (Fig. 14.8). The management and cropping of the polythene tunnels used in the Seed Production Unit of the National Vegetable Research Station, UK, was described by Faulkner (1983).

CULTIVAR MAINTENANCE

During the course of cultivar maintenance or multiplication to increase the quantity of available seed, it is necessary to ensure that the product will be 'true to type'. This trueness to type infers that the plants grown from the new seed-lot do not differ significantly from the cultivar description. The loss or deterioration of a cultivar's unique character is sometimes referred to as loss of trueness to type or 'running out'. Trueness to type may be referred to as the 'cultivar purity standard' and certifying authorities clearly state the maximum permissible level of distinguishable off-types. Only seed stocks conforming to a varietal description should be certified.

The crop is inspected at stages to ensure that any undesirable material is removed as far as possible before seeding takes place. This process of removing 'off-types' is frequently referred to as 'roguing'. The generally accepted roguing stages for individual crops are given in Chapters 7–16 although plant breeders, who are responsible for the maintenance of their own cultivars, or certification authorities responsible for specifying crop inspections for certification purposes will produce their own criteria for the different stages.

Generally the level of uniformity is lower in cross-pollinated crops such as European spinach (*Spinacea oleracea* L.) and the different cultivated types of *Brassica oleracea* L. (although summer cauliflower is a notable exception (Watts 1963)), than in the predominantly self-pollinated crops such as lettuce and tomato. The reason for the larger variation of the cross-pollinated crops is that they are based on a wider range of genotypes and the cross-fertilization between the different types maintains the heterozygosity.

Positive selection

This system is also referred to as 'mass selection'. The plants are normally selected to a very high standard while still in their vegetative stage. The percentage of plants selected by this method for seed production depends on the observed degree of variation in the crop, but the selection of approximately 10 per cent from the overall population is seldom exceeded. It is the method most commonly used for maintaining breeders' pre-basic and basic seed stocks. In biennial crops it is customary to lift the plants and re-plant them in a block. In some cases this is done in a structure, cage or relatively well isolated area. For root and bulb crops such as carrots and onions this selection can be done at the end of their first season. Although this method is extremely useful for crops in Cucurbitaceae, such as cucumbers and watermelons, in practice the selected plants have to remain *in situ* and there may be relatively long distances between selected plants. The seeds from a positive selection are harvested in bulk unless 'progeny testing' is to be done (see below).

Selected plants are usually marked by a cane and may be inspected more than once depending on when the desirable characters are best observed. It is important that the selected plants are moved to their seeding quarters before the unselected plants start flowering, or if the selected plants are to remain *in situ* then the unselected plants are removed well before they commence flowering.

Negative selection

This system is frequently referred to as roguing. It is the method most frequently used for open-pollinated crops in the final stages of multiplication. Only the plants regarded as atypical are removed from a population, the remainder normally stay *in situ* for seeding. Although the percentage of plants 'rogued out' differs according to the species' method of pollination and purity of the stock, the amount of plants removed may be approximately 20 per cent. If 20 per cent are removed by negative selection, this could also be regarded as a positive selection of 80 per cent thus the system of negative selection provides a much lower selection pressure than the positive selection method.

Positive selection and vegetative propagation

This modification of positive selection has been used for some of the brassica crops including cauliflowers and Brussels sprouts. Vegetative propagules are taken from individual plants, rooted and the resulting plants are subsequently mass–pollinated and seeded. Vegetative propagation methods used for cauliflower are described in Chapter 9. With the advances made with micro-propagation it is likely that this method will become increasingly important in the maintenance of some cultivars of a wider range of crops.

Progeny testing

This technique is widely used for testing the material obtained for breeders' stocks but it can be used at any stage of multiplication. It is especially useful for crops which are cross-pollinating but can also be used for inbreeders. Johnson (1958) recommended progeny testing for improving market yields of Brussels sprouts. Since that time the majority of Brussels sprout cultivars are F_1 hybrids but the progeny testing technique has been used more recently

for the production of spring and winter cabbage cultivars (Faulkner 1983).
It is also widely used for improving stocks of the vegetables in Cucurbitaceae
including melons and watermelons.

Selected plants are open-pollinated but the seed from each selection is harvested separately. A small sample from each seed-lot is grown on as separate
progeny. The performance of each progeny is assessed and the main seed-lots
from which the satisfactory progeny lines were grown are bulked together
and used for further seed production. The individual seed-lots from which
the undesirable progenies were grown are discarded (Fig. 3.4).

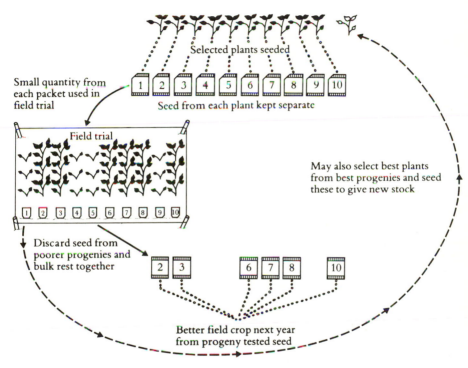

Fig. 3.4   Diagrammatic representation of progeny testing (from Faulkner 1983).

The intensity of roguing or selection during seed multiplication can have
a considerable effect on the genetical quality of successive generations. Johnson and Haigh (1966) demonstrated that if selection standards were relaxed
in three successive generations of Brussels sprouts (a cross-pollinating crop),
there could be an increase in the proportion of off-types. Dowker *et al.*
(1971), investigating the effect of selection on bolting resistance and quality
of red beet, found that the type of selection imposed during multiplication
and the season during which the selections were made produced stocks with
bolting differences in subsequent generations.

There are different methods of ensuring that the most desirable plants in
a population are used for the maintenance of seed stocks; these are 'positive
selection' and 'negative selection'. In addition the technique of progeny testing can be used to ensure that only plants with high genetical quality are used
to contribute to the following generation.

The scrutiny of plants to be retained from a population grown for seed production not only includes checking for trueness to type but the removal of any other plants showing undesirable features. In practice therefore the roguing operation includes removal of diseased plants, with particular attention to seed-borne pests and pathogens, cross-compatible weeds or wild species and other crop plants which are likely to contaminate the genetical or mechanical purity of the seed-lot.

THE PRACTICAL IMPROVEMENT OF ROGUING EFFICIENCY

The roguing operation is the basis for ensuring that the seed-lot is of the highest possible genetical quality. The following points are therefore emphasized in order to achieve an efficient operation.

The cultivation method should ensure that plants can be observed individually whatever the *optimum* crop density for the species. If field operations such as thinning are not done efficiently, the identification of any smaller but undesirable plants will be obstructed by larger, normal plants. The roguing operation should be planned so that each worker in the roguing gang can view his assignment and that there is no 'underlap' of unseen plants between the allocation of any two workers.

It is extremely important that the off-type or rogue plants are removed in their entirety so that they do not subsequently flower.

It is important to consider the position of the sun and the effect of high temperatures on the plants. Bright sunlight reduces the possibility of detecting colour differences, and the work should be done relatively early in the day before any effect of wilting masks morphological characters and the brighter sunlight masks any subtle characters. Ideally the sun should be coming over the shoulder of the person roguing rather than shining into the eyes. The roguing should be done when the crop reaches the appropriate stage and any delays should be avoided.

ISOLATION

One major factor during the course of seed production is to ensure that the possibility of cross-pollination between different cross-compatible plots or fields is minimized. This is achieved either by ensuring that crops which are likely to cross-pollinate are not flowering at the same time (i.e. isolation by time) or that they are isolated by distance.

In addition to the question of undesirable cross-pollination, adequate isolation also assists in avoiding admixture during harvesting and the transmission of pests and pathogens from alternative host crops.

Tolerance limits to genetical contamination

The maximum permissible or acceptable contamination resulting from un-desirable cross-pollination will depend on the species and on the class of seed to be produced. It follows that cross-pollinated crops will have a higher de-gree of variation than a self-pollinated crop. While it may be thought desir-able to reduce pollen contamination to zero, the amount of permissible contamination will vary with the species and the purpose for which the seed stock is intended. Even if it were possible to exclude pollen contamination completely it would not be possible to have a tolerance limit lower than the species' mutation rate (Bateman 1946).

The higher the class of seed, the lower the acceptable number of off-types and these levels are specified by certification authorities according to the crop and seed class.

Isolation in time

This type of isolation is possible within individual farms or multiplication stations. In the planning stage it is arranged that the cross-compatible crops are grown in successive years or seasons. This principle is easier to achieve in areas of the world where the climate allows two successive crops to be grown in a year. Seed multiplication stations responsible for the multipli-cation of relatively few cultivars can plan production so that no two cross-compatible cultivars are multiplied simultaneously.

Despite isolation in time there still remains the need to ensure that crops are isolated by distance.

Isolation by distance

This type of isolation is based on the concept that if a seed crop is sufficiently distant from any other cross-compatible crop then adverse pollen contami-nation will be negligible. In practice it is impossible to prevent completely 'foreign' pollen reaching a crop because the wind can carry pollen grains or pollinating insects relatively long distances.

Regulations or recommendations for isolation distances of specific crops take into account the method of pollination (i.e. whether the species is pre-dominantly self- or cross-pollinated) and the pollen vector (i.e. insect or wind). In some countries (e.g. the Netherlands) the minimum isolation dis-tance between different groups or types of cultivars of the same species is greater than for cultivars of a similar type. For example, the minimum iso-lation distance between runner bean cultivars with different flower colours is greater than that for cultivars with the same flower colour. The specified distances are also greater for classes of seeds to be used for further multipli-cation than for those to be distributed to growers for production of a market crop. The generally accepted isolation distances for specific crops are given in Chapters 7–16 but these are only guidelines as the regulations vary between countries. Much of the experimental work investigating optimal isolation distances has been done in temperate regions and factors such as topography, natural barriers, prevailing wind and insect populations can influence the efficiency of isolation distances.

In practice there are several possible sources of pollen which can contaminate a seed crop during anthesis. In addition to other seed crops pollen can originate from cultivar trials, private gardens, volunteer plants, early bolters in cross-compatible crops and wild or escape species.

ZONING

In addition to the primary isolation requirement for a seed crop there are in some countries or areas of countries zoning schemes which control the species to be grown either for market or as seed crops. In principle the specification of what is allowed to be grown in a given zone ensures that cross-compatible species or types of related crops do not freely cross-pollinate. For example, there are different types of *Beta* species, including mangel, fodder beet, sugar beet and red beet. By allowing only one of these types to be grown in a specified area or zone the chances of highly undesirable cross-pollination between the types will be greatly reduced. Other horticultural crop groups for which these arrangements are organized in the UK include *Brassica campestris* (turnip), *Brassica oleracea* (cole crops) and *Allium* species (Anon. 1981). Zoning regulations may also call for the registration of any seed or market crop in a specified area regardless of the purpose for which it is to be grown. In the USA sweet corn seed is produced in Idaho where it is isolated from maize with which it freely cross-pollinates (Delouche 1980).

The use of zones by either voluntary or compulsory schemes is especially useful for wind-pollinated crops such as red beet and sweet corn. The schemes are even more important where the commercial crop production for consumption has to flower before a marketable crop is produced, as in the case of *Zea mays* L.; it is also extremely important where the seed crop is normally biennial but adjacent crops, not intended for seed production, bolt in their first year as sometimes occurs in the different types of *Beta vulgaris* L.

DISCARD STRIP TECHNIQUE

According to Dark (1971) the pollen concentration in the air over a field of a wind-pollinated crop increases from the windward edge downwind with a tendency to decline again at the lee edge. Therefore during the period of anthesis when the wind would have been blowing from every direction, a strip around the field's perimeter would have received relatively little of the crop's own pollen and there would have been maximum concentration in the centre. The marginal strip is especially important in the production of genetically pure seeds. When a cloud of contaminant pollen passes over the field a small number of pollen grains will drop out at random. Those falling over the centre of the plot will compete with the relatively high concentration of the crop's own pollen and have a lower chance of fertilizing, whereas those which fall on marginal areas will not have so much competition and will therefore have a higher chance of fertilization. Thus if the seeds from a 5 m wide strip around the perimeter of the plot are harvested separately they can either be destroyed or placed in a lower category according to the genetical quality tested by growing a sample. The bulk of the seed will come from the inner area and can be kept as a separate seed-lot.

FIELD OR PLOT SHAPE

Most pollen contamination of either wind- or insect-pollinated crops occurs around the perimeter of the plot or field. Therefore if the area of each crop is kept as near as possible to a square then fewer seeds are likely to have been produced as a result of undesirable pollen and the minimum amount of seed is involved if the discard strip technique is applied.

TRIAL GROUNDS AND GROWING ON TESTS

The types of trials discussed in Chapter 1 included the cultivar trials, testing for distinctness, uniformity and stability (DUS) and trials by plant breeders to evaluate material in the course of development. In addition to these trials seed companies will normally grow on a sample of each new seed-lot as a check on its genetical purity. This growing on test (sometimes referred to as 'growing out') is in addition to the standard germination test (discussed in Chapter 6) and any other trials for which samples may have to be submitted. Many companies also include commercial samples from their competitors, especially of new cultivars. In this way a seed house not only monitors the growing on quality of its seed stocks but also is able to see the cultivars marketed by other companies and new releases from research institutes.

PLANT PRODUCTION

The agronomic methods and systems used to obtain a satisfactory plant stand for seed production are generally the same as for the production of fresh vegetables and either direct drilling or transplanting are used according to the crop species and local conditions. The methods for individual crops are described in Chapters 7–16.

There are two terms which are commonly used by vegetable seed producers according to the method of production, viz. 'seed to seed' and 'root to seed'. These terms are frequently used when referring to biennial root and bulb crops. 'Seed to seed' refers to a crop which is sown direct in its final quarters and the plants remain *in situ* at the end of their first season. With this system in which the plants overwinter without being transplanted, it is not possible for selection or roguing of root characters and it is therefore used only for the final multiplication stage of crops such as carrots and onions and should never be used for basic seed production.

The term 'root to seed' is generally used for biennial crops when the plants are to be lifted at the end of their first season and re-planted after discarding the undesirable material. In some areas the overwintering material is stored between lifting and re-planting.

Some root crops, for example red beet, are sown later in their first season and are transplanted as relatively small plants either before or after the winter. Plants produced for this method are frequently referred to as 'stecklings' and while they are of insufficient maturity when transplanted for all their characters to be assessed, some degree of verification of characters can be made depending on the species.

ADMIXTURE

An important factor to emphasize in seed production is the prevention of admixture. This is the contamination of one seed-lot with another and is especially hazardous to satisfactory cultivar maintenance if closely related stocks are mixed. Therefore care must be taken throughout the sequence of operations to ensure that admixture does not occur. Sources of this type of contamination include seed drills, processing machines and human error.

MOTHER PLANT ENVIRONMENT

The modification of the microclimate within seed production areas was discussed above but it is also important to ensure that the soil or substrate is not a limiting factor. Generally the same principles of crop nutrition apply in seed production as for the market crop. However, in view of the increased time required to produce a seed crop, the loss of nutrients due to leaching must be taken into account by the application of appropriate top dressings in the late stages of plant development. There is increasing evidence that the nutrient regime of the mother plant not only influences seed yield but also seed quality. For example, Browning and George (1981a, 1981b) showed the influence of NPK regimes on pea seed yield and quality.

The optimum soil pH and ratios of NPK fertilizers for specific crops, together with important micro-nutrient requirements, are given in Chapters 7–16.

THE CONTROL OF PESTS, DISEASES AND WEEDS

Generally the same control methods are used in seed production as for the production of market crops of vegetables. The range of available pesticides differs from one country to another but only approved or proven products should be used in seed production as possible adverse effects of pesticides include the inadvertent killing of pollinating insects, a modification of the seed's potential germination and a reduction in seed quality. Little work has been done on this last aspect of the effect of pesticides on seed quality but Olympio (1980) showed that while some herbicides did not affect seed yield or quality in lettuce and tomato, trifluralin at the recommended rate increased tomato seed yield whereas diphenamid decreased seed quality. Propyzamide applied to lettuce at or after anthesis reduced the seed yield. Faulkner (1983) listed the approved pesticide products suitable for use in the seed production of some vegetables in the UK.

Pre-sowing seed treatments

There are several pre-sowing treatments which are used for vegetable seeds, including the application of pesticides for the control of seed- or soil-borne pathogens, modification of seed shape or size and pre-germination before sowing.

Seed treatments for the control of pathogens

The range of vegetable seed treatments and their methods of application for the control of fungi, bacteria and insect pests has been reviewed by Maude (1978). The available treatments range from the application of chemicals to the seed as dusts or slurries to the application of heat via hot water, dry heat or steam-air mixtures. The methods for the application of pesticides to seeds were reviewed by Jeffs and Tuppen (1978). The main seed-borne pathogens with the common names of the diseases they cause are given in Chapters 7–16 for the crops discussed.

Fumigation

Seed-borne eelworms are controlled in onion and leek seeds by fumigation with methyl bromide. Fumigation is done in sealed containers and extreme care must be taken during the operation because of the hazard to the operators and other persons. Baker (1972) reported that 25 mg/litre is used at a temperature of 24 °C for 24 hours.

PELLETING AND COATING

Pelleting

Pelleting facilitates the manual and mechanical handling of otherwise small or awkwardly shaped seeds. Individual seeds are encased in an inert material such as montmorillonite clay which may also contain appropriate pesticides. The pellet has to be sufficiently hard to allow handling but capable of breaking down on contact with soil moisture. The technology for pelleting vegetable seeds has been gained largely from developments with pelleting sugar beet seed (Longden 1975). The actual pelleting is usually done by specialist companies on a contract from the seed company. Tonkin (1979) reviewed the implications of seed pelleting with special reference to seed quality and testing.

Coating

Seed coating is a technique by which additives such as pesticides, nutrients or nitrifying bacteria are applied to the external surface (i.e. testa) of the seed. But in contrast to pelleting the coating conforms to the individual seed's shape and does not normally significantly modify the seed's size.

FLUID DRILLING AND SEED PRIMING

The modern vegetable production systems depend to a large extent on high quality seeds but the field environment factors can affect germination, subsequent seedling emergence and the population of established seedlings.

Two research topics which have attempted to eliminate uncontrollable field factors are the sowing of pre-germinated seeds in a protective fluid carrier (i.e. 'fluid drilling') and the technique of seed 'priming'.

The technique of fluid drilling pre-germinated seeds and the possibilities which it may offer were discussed by Salter (1978). Basically the system involves uniform seed germination, storage of the germinated seeds without root growth if soil or weather conditions prevent immediate drilling and the subsequent handling, transportation and drilling of the germinated seeds.

The priming of seed is a pre-sowing treatment which aims to have all seeds in a given lot on the 'brink' of germination prior to sowing. Thus while an initial seed-lot may contain a population of seeds with differing potential durations to germination, the technique of priming will ensure that each seed in the sample has reached the same stage of germination prior to sowing. The technique of priming, which is in two basic stages, was described by Heydecker and Gibbins (1978) and Heydecker (1978). Firstly, the seeds imbibe water under aerobic conditions using a solution of controlled water potential; materials such as polyethylene glycol have been used for this stage, which ensures that the seeds absorb almost sufficient water for radicle emergence.

Secondly, the seeds are sown surface-dried or air-dried after a period of storage.

While the successful adoption of fluid drilling or seed priming has not met earlier expectations, both techniques offer enormous potential in the future, and with further modifications and development may well become widely used commercially.

REFERENCES

Anon. (1981) *Guide to the Arrangements in North Essex for the Prevention of Injurious Cross-Pollination of Seed Crops of Allium (Onion), Beta (Beet) and Brassicas.* Ministry of Agriculture, Fisheries and Food, London.

Baker, K. F. (1972) Seed pathology. In *Seed Biology*, T. T. Kozlowski (ed.). Academic Press, New York and London.

Bateman, A. J. (1946) Genetical aspects of seed-growing, *Nature* **157** 752–5.

Browning, T. and George, R. A. T. (1981a) The effects of nitrogen and phosphorus on seed yield and composition in peas, *Plant and Soil* **61** 485–8.

Browning, T. H and George, R. A. T. (1981b) The effects of mother plant nitrogen and phosphorus nutrition on hollow heart and blanching of pea (*Pisum sativum* L.) seed. *J. Exp. Bot.* **32**(130), 1085–90.

Dark, S. O. S. (1971) Experiments on the cross-pollination of sugar beet in the field, *J. nat. Inst. Agric. Bot.* **12**, 242–266.

Delouche, J. C. (1980) Environmental effects on seed development and seed quality, *Hortscience* **15**, 13–18.

Dowker, B. D., Bowman, A. R. A. and Faulkner, G. J. (1971) The effect of selection during multiplication on the bolting resistance and internal quality of Avon early red beet, *J. Hort. Sci.* **46**(3), 307–11.

Faulkner, G. J. (1983) *Maintenance, Testing and Seed Production of Vegetable Stocks.* Vegetable Research Trust, NVRS, Wellesbourne, Warwick, pp. 62.

Gaunt, R. E. and Liew, R. S. S. (1981) Control of diseases in New Zealand broad bean seed production crops, *Acta Horticulturae* **111**, 109–12.

Heydecker, W. (1978) Primed seed for better crop establishment, *Span* **21**(1), 12–14.

Heydecker, W. and Gibbins, B. M. (1978) The 'priming' of seeds, *Acta Horticulturae* **83**, 213–23.

Hrabovszky, J. P. (1982) Crop production in developing countries in 2000. In *Seeds*, W. P. Feistritzer (ed.), pp. 29–39.

Jeffs, K. A. and Tuppen, R. J. (1978) The application of pesticides to seeds. In *CIPAC Monograph 2, Seed Treatment*, K. A. Jeffs (ed.), pp. 10–23. Heffers, Cambridge.

Johnson, A. G. (1958) How to increase top quality sprout yields, *Grower*, November 1.

Johnson, A. G. and Haigh, J. C. (1966) The effect of intensity of selection during successive generations of seed multiplication on the field performance of Brussels sprouts, *Euphytica* **15**, 365–73.

Longden, P. C. (1975) Sugar beet seed pelleting, *ADAS Quart. Rev.* **18**, 73–80.

Maude, R. B. (1978) Vegetable seed treatments. In *CIPAC Monograph 2, Seed Treatment* K. A. Jeffs (ed.), pp. 91–99. Heffers, Cambridge.

Olympio, N. S. (1980) Influence of certain pesticides on the yield and quality of lettuce, rape and tomato seeds. *Ph.D. Thesis*, University of Bath.

Salter, P. J. (1978) Fluid drilling of pre-germinated seeds: progress and possibilities, *Acta Horticulturae* **83**, 245–9.

Tonkin, J. H. B. (1979) Pelleting and other presowing treatments. *In Advances in Research and Technology of Seeds* Volume 4. J. R. Thomson (ed.), pp. 84–105.

Watts, L. E. (1963) Investigations into breeding systems of cauliflower (*Brassica oleracea* var. *botrytis* L.). I. Studies of self-incompatibility, *Euphytica* **12**, 323–340.

Webster C. C. and Wilson, P. N. (1980) *Agriculture in the Tropics.* Longman, London and New York.

FURTHER READING

Anon. (1981) *Insect Pests of Brassica Seed Crops.* Leaflet 576, Ministry of Agriculture (Publications) Alnwick, Northumberland.

Austin, R. B. (1963) Yield of onions from seed as affected by place and method of seed production, *J. Hort. Sci.* **38**, 277–85.

Austin, R. B. (1972) Effects of environment before harvesting on viability. In *Viability of Seeds*, E. H. Roberts (ed.), pp. 114–149. Chapman and Hall, London.

Bateman, A. J. (1944) Contamination in seed crops. III. Relation with isolation distance, *Heredity* **1**, 303–36.

Baxter, S. M. (1978) *Windbreaks.* MAFF, London.

Chamberlain, A. C. (1967) Cross-pollination between fields of sugar beet, *Quart. J. Roy. Meteorol. Soc.* **93**, 509–15.

Crane, M. B. and Mather, K. (1943) The natural cross-pollination of crops with particular reference to the radish, *Ann. Appl. Biol.* **30(4)**, 301–8.

Flegmann, A. W. and George, R. A. T. (1975) *Soils and Other Growth Media.* Macmillan, London.

George, R. A. T. (1980) Seed nutrition. In *Crop Seed and Soil Environment*, pp. 106–110. Ministry of Agriculture, Fisheries and Food, HMSO, London.

Haskell, G. (1943) Spatial isolation of seed crops, *Nature* **152**, 591–2.

Neergaard, P. (1977) *Seed Pathology*, Volumes 1 and 2. Macmillan, London.

Seemann, J., Chirkov, Y. I., Lomas, J. and Primault, B. (1979) *Agrometeorology.* Springer-Verlag, Berlin, Heidelberg, New York.

4 HARVESTING AND PROCESSING

The period of anthesis of many of the cultivated vegetable species is relatively long, with successive flowers on complex inflorescences, and resulting in a long period of fruit ripening and seed maturation. In many species, e.g. lettuce, *Brassica* spp. and okra (Fig. 16.3), there is a strong tendency for the earlier maturing seeds to drop before later ones have developed. This loss of seeds before harvesting takes place is often referred to as 'shattering' or 'shedding'. There has been an active interest in the use of polyvinyl acetate sprayed on to seed crops to prevent shattering. The material acts as a glue which dries on the plant and prevents shedding while subsequent seeds develop (Williams 1977).

Other sources of seed loss before harvesting include birds, small rodents and inclement weather. There is a wide range of bird scaring devices used throughout the world to prevent loss and they include audio systems (such as methane 'bangers' or boys with rattles), scarecrows and aerial balloons. Netting is successful for small areas of basic seed but is impractical over large areas.

LODGING

This term is used to describe the collapse of a crop before cutting or harvesting. Seed crops are especially vulnerable after the plants have bolted and there is the extra weight of the inflorescences or seed heads at the top of the plants.

In addition to the susceptibility of specific crops, such as lettuces, to lodge after bolting, several cultural and environmental factors can contribute to the incidence of lodging; these include wind, high nitrogen regimes during earlier plant development, heavy rain which can either weigh plants down or reduce efficiency of root anchorage, and straying animals.

Once lodging has occurred the plants tend to deteriorate and, depending on their development stage, may not regain their vertical posture. This results in a poor micro-environment in the crop canopy, and in relatively wet seasons or areas the seed quality may deteriorate with a reduction in yield and germination.

STAGE OF HARVESTING

The effects of harvesting stages of peas and carrots have been reported by Biddle (1981) and Gray (1983) respectively. The work with peas emphasized the need for pea seed moisture content to be between 30 and 44 per cent for threshing in a pea viner or approximately 26 per cent when combine harvested. Gray's work with carrots clearly showed a relationship between embryo size not only with position on the plant but with seed maturity. The effects of carrot umbel order and harvest date on seed variability and seedling performance was reported by Gray and Steckel (1983). While the above examples of effects of harvesting stage on seed quality demonstrate the complexity of the topic, there is little published work on these aspects in relation to other vegetable seed crops and there is scope for further research.

In addition to the seeds' development and ripening rate characteristic for the species, ripening is accelerated by relatively high temperatures, a low soil moisture level and a low relative humidity. Conversely the rate is reduced by the reciprocal of these factors. The effects of environment before harvesting on seed viability was reviewed by Austin (1972). The ripening process is interrupted if the seeds are harvested too early and seed quality may be adversely affected.

Generally the later the harvest the higher the seed yield but as shown in Fig. 4.1 losses increase as harvesting is delayed once the optimum percentage of seeds reach maturity. Thus for any individual crop the ideal harvest time is immediately before the loss of mature seeds exceeds the amount of seeds yet to reach maturity.

The incidence of shedding from ripe material is increased during dry weather. Crops which are particularly prone to shedding while being cut should be handled at times of comparatively high relative humidity. In arid areas this can include early in the day when the effects of overnight dew are still effective, after rain or even following irrigation.

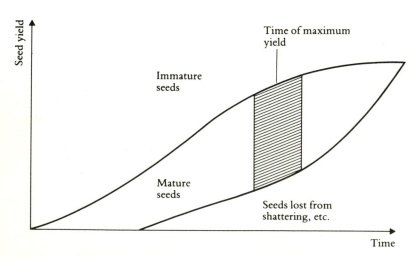

Fig. 4.1   Schematic representation of the interaction of seed maturity, potential yield and loss from shattering and birds.

Although seeds which are produced in wet or fleshy fruits such as cucurbits, peppers and tomatoes are not normally subject to all the effects of environment as are the dry seeded vegetable species, it is generally accepted that the seeds should be allowed to develop fully in the fruit before extraction. Ideally the fruit should remain attached to the mother plant but work by Cochran (1943) showed that the stage of fruit maturity at time of harvest had a significant effect on potential germination of *Capsicum frutescens* (pimiento pepper): seeds extracted from thirty-day-old fruits gave approximately 5 per cent seedling emergence when tested in compost, but fruit picked at the same age and stage which were kept for a further thirty days before seed extraction produced seeds which gave a seedling emergence of 95 per cent.

TYPES OF MATERIAL TO BE HARVESTED

The types of vegetable seed material to be harvested can be broadly classified into one of three groups, viz.:

1. Dry seeds (e.g. brassicas, legumes and onion).
2. Fleshy fruits which are dried before seed extraction (e.g. chillies and okra).
3. Wet fleshy fruits (e.g. cucumbers, melons and tomatoes).

In some crops the method of harvesting and extraction depends on the scale of operation, the stage of multiplication and on any links with food processing. For example, pepper seeds are extracted from the wet fruits if large quantities are produced in conjunction with a dehydration plant (Fig. 4.2), but if only small quantities are produced the seeds are extracted from relatively dry fruits after fully ripening.

HARVESTING DRY SEEDS

The methods of harvesting dry seeds include cutting off individual seed heads (Fig. 13.6); cutting the majority of the plant and leaving the material in windrows to dry further before the seed is extracted; or the use of a combine harvester which extracts the seed from the seed heads and separates it from the remainder of the plant debris (i.e. the 'straw') in one operation. In the first two methods (i.e. removal of individual seed heads and cutting for further drying) the seed is subsequently separated from the 'straw' by threshing.

Harvesting by hand

Hand harvesting is still done for very high value seeds, when the total area to be harvested is very small (e.g. breeders' or basic seed) or in areas where

Fig. 4.2   Harvesting pepper fruits in California, USA, for seed extraction in conjunction with a dehydration factory.

there is adequate hand labour available. Seed heads, dried fruits or other forms of modified inflorescences containing the mature seeds (e.g. sweet corn cobs and leek flower heads) are picked or cut with knives or secateurs into baskets or other suitable containers. In some crops which are cut by hand a larger part of the plant is removed with the seed heads. This is achieved, for example with radish, lettuce and brassicas, by the use of knives, matchets, sickles or in the case of some crops, such as peas, by pulling up the whole plant.

Hand-harvested material is usually either dried further on tarpaulins or other suitable sheets, or placed in suitable buildings or structures on clean concrete floors or in airy racks or boxes.

Desiccants

In some seed crops it is necessary to dry-up the plant material when the seed is approaching maturity. This facilitates mechanical harvesting of the seed by reducing the amount of plant debris, increases the rate of seed drying, brings all the seed material to a closer range of moisture content, avoids loss from windrows and will also to some extent control weeds. Materials which are used for this purpose are known as desiccants of which an example is diquat. However, several other materials have been used in trials of vegetable seed production. The effects of desiccants on yields and quality of carrot and chicory seeds are reported by Lovato and Montanari (1981) and Montanari and Lovato (1981), and the effects on *Phaseolus vulgaris* by Hole and Hardwick (1978).

Mechanized cutting

When the cutting operation is mechanized the material is either left *in situ* by a machine with a cutter bar or a machine which cuts and places the material in windrows. Machines capable of this operation have a draper belt in addition to the cutter bar and the cut material is carried under the machine and deposited in a swath. The swaths can be either turned into windrows or left where they are to dry according to the density of the material and the rate at which it will dry in the field.

Combine harvesting

This joint operation is done by combine harvesters which cut and thresh the material in one operation. Material which has already been cut can be picked up from windrows and combine harvesters can also be used as stationary threshers (see below).

Threshing

Dry seeds are removed from the mother plant material by flailing, beating or rolling the material. It is important to ensure that unnecessary fragmentation of plant material does not produce debris which is either difficult or costly to separate from the seed sample by subsequent processing; it is also extremely important to avoid damaging the seeds.

Hand threshing

Hand threshing is a relatively cheap method for small seed-lots and is still used in some countries for large seed-lots where labour is cheap. Several hand methods are available including rubbing, beating the material against a wall or the ground, or flailing.

Rubbing seed materials with a presser in an open-ended trough lined with ribbed rubber is very suitable for material in pod-like structures such as brassicas and radish (Fig. 4.3).

Seeds which have been hand threshed are usually still mixed with the plant debris and further separation is done by winnowing or sieving.

Fig. 4.3   Method of dry seed extraction for small samples.

Machine threshing

The main feature of threshing machines is a revolving cylinder in a concave (Fig. 4.4). The cylinder is driven by a motor or engine and is capable of reaching 1200–1500 revolutions per minute (rpm). While the potential of these high rpm's is suitable for agricultural grain, speeds of 1100 rpm are used for most small seeded vegetable species, and as low as 700 rpm for large seeded legumes.

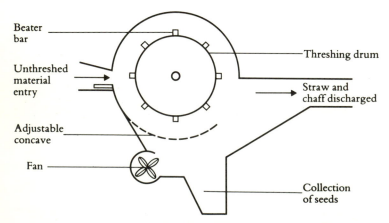

Fig. 4.4   Diagram of a thresher showing position of concave.

Both the cylinder and the concave may have either spiked steel teeth, angle bars, rasps or rubber bars, although threshers with bars are more frequently used for vegetable seeds than models with teeth. The concave opening is adjusted to allow the free passage of seeds which are collected in a container, usually below the drum. Some threshers have interchangeable concaves for use with large seeded legumes.

Some types of thresher incorporate sieving, screening or aspirating components to assist in the initial separation of seeds from plant debris. These modifications include the possibility of adjusting cylinder speed, cylinder clearance, concave mesh, air flow and screens. These refinements are essential for dealing with a range of vegetable species.

Damage to seed during threshing

The possibility of surface and sub-surface damage to seeds is increased during threshing if the cylinder speed is too fast, the cylinder clearance is too narrow or the mesh of the concaves is too small.

SEED EXTRACTION FROM WET OR FLESHY FRUITS

Extraction of seeds from wet or fleshy fruits is either done by hand or by specially designed machines. The specific extraction methods, appropriate machines and special cleaning requirements are described for Cucurbitaceae in Chapter 10 and Solanaceae in Chapter 12.

SEED PROCESSING

The term 'seed processing' is used by the seed industry to include a wide range of operations to improve or 'upgrade' seed-lots after threshing or extraction. The objectives of processing may include removal of a wide range of materials including plant debris, non-plant material (e.g. soil or stones), seeds of other crops and weeds, seed appendages which would otherwise interfere with the free running of the seeds, or damaged and discoloured seeds and seeds which are outside the accepted size or density tolerance of the seed-lot.

Principles of seed processing

The separation of seeds from other materials is based on physical differences such as relative size, shape, length, density, surface texture, colour, affinity to liquids or relative conductivity. In practice the first five of these characters are widely used in the design of equipment and the last three are used only in very specialized equipment. The specific machine used will depend on the crop, stage of processing and any special cleaning problems.

The range of operations can be subdivided into four basic groups, viz. winnowing, pre-cleaning (often referred to as 'conditioning' or 'scalping'), basic cleaning and separation (or 'upgrading').

Winnowing

After threshing dried seeds can be separated from less dense debris by winnowing; the operation can be done by hand or machine.

Pre-cleaning ('scalping')

During pre-cleaning the bulk of plant debris and any other non-seed materials are separated by vibrating or rotating sieves. In some species seed clusters are also separated during pre-cleaning. The pre-cleaning machines usually have an air flow to remove materials lighter than the seed. Seed-lots are usually pre-cleaned before drying.

Basic cleaning

The main cleaning operation is generally referred to as basic cleaning. During this stage all materials should be removed from the seed crop with the possible exception of contaminants which require a special separation process. The simplest form of basic cleaning is the use of sieves (Fig. 4.5) which separate on size. Modern seed cleaners also incorporate motorized fans which force air through the cleaning chamber or draw air out of it. These are generally referred to as 'air-screen' machines and have at least two vibrating screens (Fig. 4.6).

Separation and upgrading

These are normally the final processes which improve the mechanical purity of the seed-lot and may be done in order to remove a specific contaminant or appendages from the crop seed (in this latter case the operation is usually performed early on in the processing sequence).

Spiral separator

The spiral separator, which has a minimum of two spirals around the vertical axis, is used for separating non-spherical or irregular-shaped seeds from a round-seeded species, for example to separate *Brassica* seeds from other material and broken seeds.

Disc and cylinder separators

Generally in the modern seed processing industry these types of separator have replaced the spiral separators. They operate on the principle that one fraction of a seed-lot is picked up in small depressions in a disc or cylinder

Fig. 4.5   Using sieves to separate broad bean seeds from plant debris.

Fig. 4.6   Screen and air cleaner, showing removable screens (by courtesy of Damas, Denmark).

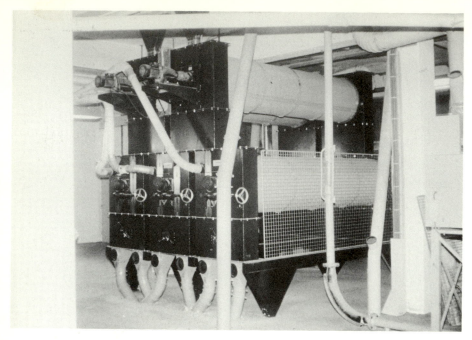

Fig. 4.7   Indented cylinder machine, with bank of cylinders and built in
          distributors (by courtesy of Damas, Denmark).

while the other fraction remains loose and is therefore separated (Fig. 4.7).
An example of their use is to separate pieces of stem from lettuce seeds and
for grading seeds of some crops. A range of discs or cylinders is available for
interchanging in order to cope with material of different sizes.

Gravity separator or gravity table
The gravity separator (Fig. 4.8) separates by specific gravity good seeds from
other materials such as seeds which are mechanically damaged, diseased,
light-weight, sterile or insect damaged. The machine is also used for grading.
The seeds are fed on to the vibrating and sloping deck above a plenum chamber and the air current separates the lighter material. The fractions are collected separately. The separation of fractions into horizontal layers is usually
referred to as stratification.

Magnetic separator
This machine is used for separation and depends on the differing surface
characters between the two fractions. The seed-lot to be upgraded is first
treated with fine iron dust which adheres to the rough coated fraction (e.g.
seeds of the common weed *Galium aparine* in a sample of radish seeds). The
material is then passed over magnetized rollers and the clean seed collected
at the side while the magnetized material is brushed off separately.

Debearder
The debearder is essentially a drum with horizontal arms rotating inside. The
arms rub the seeds against the drum and thereby remove appendages

Fig. 4.8   Gravity separator (by courtesy of Damas, Denmark).

(Fig. 13.7). This is an essential operation with carrot seed and is therefore done early on in the processing sequence to facilitate the free flow of the seeds.

Electronic colour separator
These machines are used for the separation of 'off-colour' types from a seed-lot and are frequently used for beans and peas. The seeds are fed by belt, gravity or roller past a photo-electric cell which triggers a jet of air to remove the off-colour seeds from the main seed-lot. Both monochromatic and bi-chromatic instruments are available to enable different colours as well as different intensities of the same colour to be detected.

Precision air classifier
This type of machine separates materials with a range of sizes and specific gravities by floating them through a rising air-stream. The machine works on the principle of differential specific gravities between fractions which can be separated by fine adjustments to the machine.

Managment of machines

The range of machines described above demonstrates the different principles involved in seed processing. The reader is referred to 'Further Reading' at the end of this chapter for a more complete range and detailed accounts of their operation.

Cleaning machines

It is extremely important that machines are cleaned between seed-lots. This is normally done by vacuum cleaning the removable parts such as screens and also cleaning out the interior areas. Floors and other areas around the processing machines must be kept clean and free from debris.

Seed grading

Modern vegetable production systems are based on crop uniformity and a primary contribution to this is the precision drilling of seeds. Precision drills generally rely on the grading of seeds and the seed merchant therefore has to provide farmers with seeds of specific grades. The seed trade and seed drill manufacturers have agreed on a letter coding for each of the range of seed sizes (Table 4.1).

SEED DRYING

The natural moisture content of seeds is gradually reduced as they develop and ripen. By the time it is separated from the mother plant the seed's mois-

Table 4.1   Key to code of seed grades showing seed size

Letter code	Seed size (mm)	Letter code	Seed size (mm)	Letter code	Seed size (mm)
A	0–0.25	J	2.00–2.25	S	4.00–4.25
B	0.25–0.50	K	2.25–2.50	T	4.25–4.50
C	0.50–0.75	L	2.50–2.75	U	4.50–4.75
D	0.75–1.00	M	2.75–3.00	V	4.75–5.00
E	1.00–1.25	N	3.00–3.25	W	5.00–5.25
F	1.25–1.50	P	3.25–3.50	X	5.25–5.50
G	1.50–1.75	Q	3.50–3.75	Y	5.50–5.75
H	1.75–2.00	R	3.75–4.00	Z	5.75–6.00

ture content is below 50 per cent and thereafter the moisture content is in equilibrium with the storage atmosphere.

It is frequently necessary to dry seeds when they first arrive at the processing plant from the field, after extraction from fruit or possibly after processing but before storage and packaging. The artificial drying of unthreshed seed crops was discussed by Sparenberg (1963), who reported that the depth of pea or bean material to be dried by air at 350 m³ per square metre of floor area should not exceed 2 m but for radish it could be up to 4 m when the unthreshed moisture content was at 30–45 per cent. This work demonstrated the importance of not having an excessive depth of plant material during pre-threshing drying.

Equipment for seed drying

Artificial drying systems can be classified into two main groups, viz. batch and continuous systems.

Batch drying systems

These include horizontal tray (Fig. 4.9), vertical single layer, vertical double layer and cylindrical bin with a central duct. The horizontal tray type is especially useful for dealing with different batches of seed simultaneously.

Rotary paddle hot air batch drier

The rotary paddle driers are used for the drying of tomato, pepper, eggplant and cucurbit seeds which have been extracted from fruit by a wet extraction process (Fig. 4.10).

The wet seed is placed over a finely perforated surface and dried by a current of hot air which blows through the perforations and bed of seed from below. As the hot air passes between the seed the moisture is removed to the outside air. Meanwhile, a rotary paddle progresses along the top of the apparatus, turning the seed in the bed. The bed of seed can be up to about 5 cm deep.

Advantages of the rotary paddle driers for drying seeds immediately after extraction from wet fruit are that they can be used for relatively small seed-

Fig. 4.9   Batch drier.

Fig. 4.10   Rotary paddle hot air batch drier.

lots and the temperature control is better than other batch driers. This is important when the seed is relatively wet at the start of the drying process.

For the wet-extracted solanaceous and cucurbit seeds the air temperature is controlled between 37 and 40 °C at the start, but when the moisture content begins to lower, the air temperature is reduced to 32–35 °C.

The total time for drying a batch of wet-extracted seed depends on the crop, but it is approximately 8 hours for eggplant and pepper, 10 hours for cucumbers and tomatoes and from 12 to 16 hours for the larger seeded cucurbits such as squashes, depending on the cultivar's seed size.

When the drying of the seed is thought to be complete, the hot air supply and the rotary paddle are stopped. Samples of seed are taken for a moisture content determination. If satisfactory the seed is then canned or taken for storage, but if necessary it is dried further and the moisture determination repeated. Experienced workers can usually judge when the seed's moisture level has been reduced to the required level but checks should always be made with a laboratory moisture determination.

Continuous drying systems

These systems are suitable for relatively large quantities of seed. There are two designs based on either a vertical seed flow or a horizontal bed on a perforated belt. Kreyger (1963) described the different types of seed drying equipment used in Europe and Barre (1963) described those of importance in North America.

Drying rate and temperatures

The rate at which a seed-lot can be dried artificially depends on the 'packing character' of the species and the initial moisture content of the sample. Lettuce and cucurbits are generally considered to be relatively quick driers; carrot, red beet and tomato medium driers; but legumes, brassicas, onion, leeks and sweet corn are slow driers. Generally at 43.5 °C 0.3 per cent of seed moisture content is removed per hour by an air flow rate of 4 M^3 per cubic metre of seed. Relatively slow air flow rates are cheaper to operate but fast air flow rates prevent deterioration and allow a greater throughput with the equipment. For open storage starchy seeds are reduced to a moisture content of 12 per cent and oily seeds to 9 per cent. The moisture level for seeds to be stored in sealed containers is less than these figures and is discussed in Chapter 5 and figures for sealed storage of specific crops are given in Table 5.3.

Determination of seed moisture content

The most satisfactory method of determining the moisture content of a seed-lot is the method described by ISTA (1976). This is based on an oven drying method and the reduction in weight is calculated from the weight before drying and expressed as a percentage moisture of the sample.

There are several electric meters available and while they are extremely useful as a guide for quick determinations, they are generally less reliable than the oven method and in addition require a correction factor for each species.

REFERENCES

Austin, R. B. (1972). Effects of environment before harvesting on viability. In *Viability of Seeds*, E. H. Roberts (ed.), pp. 114–49. Chapman and Hall, London and Syracuse Univ. Press. Syracuse, New York.

Barre, H. J. (1963) Important equipment for drying seeds in North America, *Proc. Int. Seed Test. Ass.* 28(4), 815–26.

Biddle, A. J. (1981). Harvesting damage in pea seed and its influence on vigour, *Acta Horticulturae* III, 243–7.

Cochran, H. L. (1943). Effect of stage of fruit maturity at time of harvest and method of drying on the germination of pimiento seed, *Proc. Amer. Soc. Hort. Sci.* **43**, 229–34.

Gray, D. (1983). Improving the quality of vegetable seeds, *Span* **26(1)**, 4–6.

Gray, D. and Steckel, J. R. A. (1983). Some effects of umbel order and harvest date of the carrot seed crop on seed variability and seedling performance, *J. Hort. Sci.* **58(1)**, 73–83.

Hole, C. C. and Hardwick, B. C. (1978) Chemical aids to drying seeds of beans (*Phaseolus vulgaris*) before harvest, *Ann. Appl. Biol.* **88**, 421–7.

ISTA (1976) *International Rules for Seed Testing.* International Seed Testing Association, Zurich.

Kreyger, J. (1963). Equipment of importance for seed drying in Europe, *Proc. Int. Seed Test. Ass.* **28(4)**, 793–826.

Lovato, A. and Montanari, M. (1981) The viability of carrot and chicory seed as affected by desiccant sprays, *Acta Horticulturae* **111**, 175–81.

Montanari, M. and Lovato, A. (1981). The yield and quality of carrot and chicory as effected by desiccant sprays, *Acta Horticulturae* **111**, 167–73.

Sparenberg, H. (1963) The artificial drying of unthreshed seed crops, *Proc. Int. Seed Test. Ass.* **28(4)**, 785–92.

Williams, C. M. J. (1977) Glueing inflorescences increases yield in vegetable crops. In *Proc. Australian Seeds Research Conference*. Tamworth, New South Wales.

FURTHER READING

Copeland, L. O. (1977) *Principles of Seed Science and Technology*. Burgess, Minneapolis.

Culpin, C. (1981) *Farm Machinery*. Granada, London, Toronto, Sydney, New York.

Delhove, G. E. and Philpott, W. L. (eds.) (1983) *World List of Seed Processing Equipment*, FAO, Rome.

George, R. A. T. (1980) *Technical Guidelines for Vegetable Seed Technology*, Chapters 10–13. FAO, Rome.

Gregg, B. R. (1983) Seed processing in the tropics, *Seed Sci. Technol.* **11**, 19–39.

Harmond, J. E., Brandenburg, N. R. and Klein, L. M. (1968) *Mechanical Seed Cleaning and Handling*. Agriculture Handbook No. 354. USDA, Washington.

ISTA (1963) Drying and storage number. *Proc. Int. Seed Test. Ass.* **28**(4).

ISTA (1977) Seed cleaning and processing, *Seed Sci. Technol.* **5**(2).

Thomson, J. R. (1979) *Introduction to Seed Technology*, Chapters 7, 8 and 9. Leonard Hill, Glasgow.

Vaughan, C. E., Gregg, B. R. and Delouche, J. C. (1968) *Seed Processing and Handling*. Mississippi State University.

5 STORAGE

When seed extraction and drying have been completed it is necessary to keep the seed under the best possible conditions to ensure that the maximum potential germination and other seed quality factors are maintained. Stored seeds are the primary input of a country's vegetable cropping programme and are vital links between successive crop generations. In commerce seed in store represents a significant proportion of the seed company's material assets.

REASONS FOR STORAGE

The period of storage may be relatively short, perhaps even only weeks but sometimes it is necessary to store seed for several years. There are many reasons for storing seed of specific crops beyond the next possible sowing season. In some cases it may be uneconomic to multiply each seed stock annually. Reasons for isolation in time have been discussed in an earlier chapter, and in addition the recurrent annual cost of multiplying each cultivar offered by a seed house or seed marketing organization has to be considered. Seed yield is influenced by many factors including provenance and seasonal variations during development and pre-harvest ripening, and it is not always possible to make accurate forecasts of yields; thus satisfactory storage is a useful way of ensuring that surplus seed is kept for future use. An increasing number of agricultural and horticultural crop programmes in developing countries are adopting the United Nations Food and Agriculture Organization's recommendation to maintain stored seed stocks for contingencies. This is especially important in countries with relatively harsh climates where unpredictable drought, flooding or other disasters can result in sudden seed shortages. Farmers' and growers' demands for seed of specific crops can depend to some extent on the general economy, market trends in crop demand or market successes or failures in previous seasons. Satisfactory seed storage will assist the seed industry to buffer these variations in requirements. In some instances specialist seed producers are in a different area of the world from the seed consumer, and stocks may arrive at their destination out of season. The viability of seeds can be adversely affected during this transit and subsequent time to sowing if they are not stored properly. It is also important to ensure that stored seed is protected from hazards such as insects, fungi, fire and pilfering.

All seed stocks are valuable and their value is related to the level of genetical purity and multiplication rate. Breeders' and basic seed stocks are normally stored under as ideal conditions as possible to reduce the frequency of their multiplication; this makes the maximum use of each stock and keeps their multiplication to a minimum.

The storage of seed as germplasm, a potentially valuable genetic resource, calls for very long-term storage of relatively small seed-lots. Germplasm centres, usually referred to as gene banks, use specialized storage methods which are not discussed here but which are dealt with in the specialist literature (see 'Further Reading' list).

NATURAL LONGEVITY

The longevity of seed is primarily dependent on its inherent keeping quality which varies with species. Some species, e.g. onion, *Allium cepa*, and leek, *Allium porrum*, are relatively short lived. Sweet corn, *Zea mays*, and the larger seeded *Phaseolus* spp. are intermediate, while seed of many genera of the Cucurbitaceae, e.g. *Cucurbita* spp., are relatively long lived in storage. In addition some workers have reported quite significant differences of storage potential between cultivars of the same species. The pre-storage history of a seed-lot will also have a very important influence on its subsequent storage qualities. Those factors affecting storage potential include mother plant environment during plant growth and seed development, and pre-harvesting conditions, especially the weather during the season of production.

Pre-harvest field factors

Micro- and macro-nutrient deficiencies during plant growth and development of a potential seed crop can have a major effect on seed storage potential in that initial germination potential is low and, although such extreme conditions do not normally prevail in commercial seed production, the levels of major nutrients have been shown to affect indirectly seed storage life.

During seed production many of the environmental effects in the field which predispose the seed to a relatively short storage life are especially severe in the tropics and sub-tropics but they may also occur in temperate regions. It is probably those arid regions which have adequate control over irrigation frequency and quantity which have the best pre-harvest quality control in relation to potential storage.

Field deterioration of seed prior to harvest is frequently associated with environmental factors such as high relative humidity, excessive rainfall and high temperature, and combinations of some, or all, of them result in a reduced potential storage life of seeds. Some of the deleterious effects can be successfully counteracted by operations such as mechanical harvesting and seed drying which are discussed elsewhere. The evidence for the effects of these operations on individual crops is discussed in other sections.

Pre-harvest mechanical factors

Mechanical damage to seed during operations such as harvesting, processing or drying can reduce the potential storage life because the damaged seeds lose vigour sooner than undamaged ones. In addition damaged seeds are more vulnerable to storage pests and pathogens. There are various forms of mechanical damage to seed which can occur before storage, one of which is threshing injury, e.g. cotyledon cracking of *Phaseolus vulgaris*, and seed coat abrasion of many species especially resulting from incorrect cylinder speed. Damage can also occur during processing if there is excessive moisture in the seed, especially if the seed is immature; the other extreme is over-drying which produces a brittle seed predisposed to breaking.

DEFINITION OF STORAGE PERIOD

The storage period may be relatively short, perhaps even only weeks, but it is also possible that seed-lots have to be stored for several years. The storage period of seed should be defined as the total time from seed maturity through to sowing, and although a seed-lot can undergo several operations such as cleaning and packaging, or a waiting period while held as stock by a retailer or farmer before being sold or sown, it is important that this total time from maturity through to sowing be regarded as the storage period, because it is throughout this entire period that the seed is subject to adverse influences of the storage environment. The phases or stages comprising the total possible storage duration of a seed-lot are comprised of some, or all, of the following:

1. Post maturity drying.
2. Seed extraction or threshing (outside or under cover).
3. Seed cleaning (outside or in buildings).
4. Holding storage (seed store or warehouse).
5. Packaging (usually in purpose-built structure).
6. Transit and distribution (all modes of transport and handling stages).
7. Marketing (wholesale and retail).
8. Point of sale (shop, garden centre or farm supplier).
9. Farm storage (from receipt of seed on farm or garden to sowing).

EFFECT OF STORAGE ENVIRONMENT ON SEED

The two most important environmental factors which can affect seed quality during storage are temperature and relative humidity. In practice the ambient relative humidity plays a major role, firstly because the seed's moisture level is a function of relative humidity, and secondly the incidence of storage fungi and insect pests is largely influenced by the relative humidity of the micro-climate within the seed mass.

Moisture content of seed

The seed most suited for storage has a moisture content not greater than 10 per cent of seed weight. A seed's metabolic rate is extremely low, often undetectable. When in this state it is hydrophilic and is capable of taking in water, even from the water vapour of the atmosphere. Thus, however much care has been taken to lower its moisture content by drying before storage it will quickly take up water again.

Moisture content of the storage atmosphere

The stored seeds can take up water from the atmosphere. The vapour pressure of atmospheric moisture at a given temperature and pressure is a direct function of the degree of moisture saturation or relative humidity. The amount of water held by air increases in relation to the air's temperature. For example, 1 kg of dry air at 10 °C can hold 7.6 g of water vapour; at 20 °C it can hold 14.8 g of water vapour and at 30 °C 26.4 g of water vapour.

Seeds of different species have characteristic moisture contents when in equilibrium with a specific relative humidity of the atmosphere. Examples of this are given in Table 5.1 in which the approximate moisture contents of the seeds of twenty-five different vegetable species are given for five different levels of relative humidity. Moisture content increases for each species of seed in relation to the increase in atmospheric relative humidity. Examples of a seed's moisture content under specific conditions of relative humidity are given in Table 5.2. This value is known as the 'equilibrium content', although some seed technologists prefer the term 'hygroscopic equilibrium'.

Control of relative humidity

There is a useful rule of thumb (Harrington 1959) which states that the storage life of seed is doubled for each 1 per cent decrease in seed moisture content. Several independent research workers have demonstrated the usefulness and accuracy of this rule at seed moisture contents between 6 and 16 per cent.

In practical terms, it is necessary to ensure that the ideal or a satisfactory acceptable low moisture content of the stored seed is maintained without the level significantly increasing. This is achieved in seed stores by maintaining the atmosphere at less than equilibrium with the optimum seed moisture content.

In arid regions, where the relative humidity is naturally low, this can be done without further investment in buildings or control systems; however, it must be appreciated that in some arid areas there are times in the year when the atmospheric relative humidity increases beyond an acceptable level.

In areas of high or fluctuating low and high relative humidities, it is vital for the life of the seed to control the relative humidity of the seed store.

Subsidiary effects of relative humidity

When the seed moisture is in the region of 40–60 per cent germination will

Table 5.1   Approximate moisture content (%) of vegetable seeds at 25.0 °C moisture content wet basis, in equilibrium with air at different relative humidities (after Harrington 1959)

Seed	Air relative humidity, %					
	10	20	30	45	60	75
Bean, broad	4.2	5.8	7.2	9.3	11.1	14.5
Bean, lima	4.6	6.6	7.7	9.2	11.0	13.8
Bean, snap	3.0	4.8	6.8	9.4	12.0	15.0
Beet, garden	2.1	4.0	5.8	7.6	9.4	11.2
Cabbage	3.2	4.6	5.4	6.4	7.6	9.6
Cabbage, Chinese	2.4	3.4	4.6	6.3	7.8	9.4
Carrot	4.5	5.9	6.8	7.9	9.2	11.6
Celery	5.8	7.0	7.8	9.0	10.4	12.4
Corn, sweet	3.8	5.8	7.0	9.0	10.6	12.8
Cucumber	2.6	4.3	5.6	7.1	8.4	10.1
Eggplant	3.1	4.9	6.3	8.0	9.8	11.9
Lettuce	2.8	4.2	5.1	5.9	7.1	9.6
Mustard, leaf	1.8	3.2	4.6	6.3	7.8	9.4
Okra	3.8	7.2	8.3	10.0	11.2	13.1
Onion	4.6	6.8	8.0	9.5	11.2	13.4
Onion, Welsh	3.4	5.1	6.9	9.4	11.8	14.0
Parsnip	5.0	6.1	7.0	8.2	9.5	11.2
Pea	5.4	7.3	8.6	10.1	11.9	15.0
Pepper	2.8	4.5	6.0	7.8	9.2	11.0
Radish	2.6	3.8	5.1	6.8	8.3	10.2
Spinach	4.6	6.5	7.8	9.5	11.1	13.2
Squash, winter	3.0	4.3	5.6	7.4	9.0	10.8
Tomato	3.2	5.0	6.3	7.8	9.2	11.1
Turnip	2.6	4.0	5.1	6.3	7.4	9.0
Watermelon	3.0	4.8	6.1	7.6	8.8	10.4

occur, resulting in the death of the seed's embryo if it occurs during storage. In addition to the direct effect on the seed's storage life, high moisture content has indirect effects, especially on the microflora and microfauna of the seed store environment. Most storage insect pests' activities, including reproduction, are stimulated with a seed moisture above about 8 per cent. Furthermore, the growth of fungi on seed will commence when the seed moisture content exceeds 12 per cent. At a seed moisture content of above 18–20 per cent, local heating can occur as a result of total metabolic heat of the biomass; this in turn can be responsible for a lowering of germination potential or even spontaneous combustion.

Temperature

Temperature is the other main factor of the seed storage environment which must be taken into account. The reduction of seed viability is slower at lower temperatures than at relatively high ones, and although this concept generally applies even when storage environment temperatures are below freezing,

Table 5.2   Germination percentages of high quality seed-lots of seven species at intervals during storage under three conditions.(Ambient Condition: Storage at State College, Miss., USA, in uninsulated metal warehouse. January mean temperature, 7 °C; July mean temperature, 30 °C; 226 frost free days: 1321 mm evenly distributed annual precipitation) (after Delouche, *et al.*, 1973)

Species	Storage condition	Storage period (months)					
		0	6	12	18	24	30
Allium cepa	7 °C–45% RH	96	—	96	—	94	94
	Ambient	96	90	42	—	—	—
	30 °C–75% RH	96	0	—	—	—	—
Citrullus lanatus	7 °C–45% RH	98	—	98	—	—	98
	Ambient	98	98	96	95	88	86
	30 °C–75% RH	98	93	31	—	—	—
Glycine max	7 °C–45% RH	94	—	94	—	96	94
	Ambient	94	94	85	60	42	—
	30 °C–75% RH	94	0	—	—	—	—
Lactuca sativa	7 °C–45% RH	96	—	95	—	90	86
	Ambient	96	90	82	68	—	—
	30 °C–75% RH	96	61	0	—	—	—
Phaseolus vulgaris	7 °C–45% RH	98	—	100	—	96	98
	Ambient	98	96	96	90	92	90
	30 °C–75% RH	98	88	21	0	—	—
Raphanus sativus	7 °C–45% RH	98	—	97	—	99	96
	Ambient	98	98	98	98	95	95
	30 °C–75% RH	98	80	13	—	—	—
Zea mays	7 °C–45% RH	98	—	98	—	98	98
	Ambient	98	98	96	96	85	65
	30 °C–75% RH	98	94	30	—	—	—

these lower extremes are normally used only for specialized storage of high value seed-lots such as germplasm.

Harrington's 'rule of thumb' which relates to temperature states that for each 5 °C increase from 0 to 50 °C the seed's potential life is decreased by half. Even quite short periods of high temperature can contribute to the reduction of seed longevity.

Combined temperature and relative humidity

In practice it is the combined effect of temperature and relative humidity which reduces potential viability or longevity of seed throughout its 'storage' life. Many areas of the world have periods of fluctuating temperatures coupled with periods of high relative humidity, the combined effect of which leads to a fast seed deterioration during relatively short periods of uncontrolled storage. Harrington (1972) has conveniently combined his separate rules for relative humidity and temperature, to state that 'the sum of the percentage relative humidity plus the temperature in degrees Fahrenheit should not exceed 100'.

The extent to which it is decided to control both these factors will depend on the value of the seed and the climate of the storage location. In some arid areas of the world the seed's potential germination is not reduced significantly during short periods of uncontrolled storage. This is because the natural drying of seed is satisfactory following seed maturity in the field and there is a low relative humidity during the subsequent storage period. Even relatively short-lived seed of species such as onion can be stored with little reduction in its germination capacity between seasons in many arid areas.

STORAGE ENVIRONMENT

The storage environment can be influenced in two main ways, viz. the building (or structure) in which the seed is stored, and its environment (i.e. temperature and relative humidity). It is important that the seed is prepared for storage in a purpose-built structure as quickly as possible after harvesting and a system organized so that the seed remains in the storage environment for as long as is practical prior to distribution for sowing.

Construction of seed stores

Seed stores should be designed to maximize security, to minimize fire risk, to exclude birds and rodents, and to keep the entry of insects and micro-organisms to a minimum.

The ideal building should have a raised and smooth finished reinforced concrete ground floor with a rodent-proof lip. Entry can be by removable ramp and the raised floor designed to match up with the loading level or height of delivery vehicles. The roof should be pitched and overhung to offer the best possible run-off of storm water and to provide shade and extra protection for the ventilator openings. There should also be a pitched canopy roof over the entrance chamber.

The extent to which rodent proofing, measures to deal with excessive storm water and extremely high temperatures are included as design features will depend on the location and prevailing conditions throughout the year.

A double door system with an entrance lobby, or antechamber, should be incorporated, and ideally there should be no other openings or windows except those connected to environmental control systems.

The walls and ceiling should be constructed of smooth finished stone, mortar and concrete which is lined with a moisture barrier of tar, aluminium foil or polythene. Wood should either not be used or kept to a minimum, as in time it can be attacked by rodents and other pests and presents maintenance problems. The final finish of the interior walls should be of a material which will protect the insulation from damage made by trolley, pallets or other handling and storage devices. Any ventilation openings which are incorporated must be efficiently screened to exclude insects. The overall interior finish must be smooth with electrical conduit channels and other cracks smoothly sealed. External finishes, especially the roof, should aim to minimize absorption of solar heat and to exclude water. Particular attention to these details is necessary in parts of the world subject to high temperatures and rainfall. Seed stores should be provided with adequate lighting with a low heat emission. There should be a regular cleaning and maintenance programme to ensure that all the design features of the structure are upheld during the life of the store.

Additional features of conditioned seed stores

The design and materials used in the seed store's construction should minimize the absorption of solar radiation and act as an effective vapour barrier. In some parts of the world these are the only features necessary to ensure a satisfactory storage life of seed. However, in many other areas further control of the storage environment must be achieved by air conditioning. This becomes a major requirement where the climate is such that design of the store is insufficient to ensure a satisfactory potential life for the seed.

The actual design and structure of conditioned stores follows the specifications outlined above. But where conditioning is required it is extremely important to ensure that the best use is made of insulating materials in order to achieve the maximum possible efficiency from temperature-controlling equipment and value from operating costs. This factor is continuing to increase in importance with the progressive rises in energy costs.

Vapour proofing is also important and is achieved by building a continuous polythene film sealed with bitumen or other suitable sealant into all floor, wall and ceiling areas. This vapour barrier forms a seal completely enveloping the store and is installed on the 'wet' or exterior side of the heat insulation barrier. All doors must also be fitted with gaskets and there should be an air lock at the entrance.

An adequate power supply for operating the apparatus which is to control the store's environment is necessary. Control appliances should be positioned so that the heat they emit is exhausted to the store's exterior. An important consideration is positioning of air exhausts; hot air should be emitted from the store just below the roof line, while moist air should be expelled from near ground level if conditioned separately.

Temperature control

Storage temperature can be reduced by ventilation and refrigeration in addition to insulation and structural features. Complete temperature control is expensive in any part of the world and is unlikely to be used in commerce for seeds, regardless of their value. Total temperature control by refrigeration is, however, used in the long-term storage of germplasm collections and breeders' material.

Temperature control during the storage of 'short-term' and 'carry over' seed stocks is achieved usually by ventilation in conjunction with refrigeration, the degree of which will depend on the outside temperature.

Ventilation

The ventilation of seed stores should be considered in conjunction with the relative humidity of the ambient air, because it would be more harmful to the seed to lower its storage temperature if the result is to increase the seed's moisture content. Conversely, ventilation can be used to lower the storage temperature and the seed's moisture content when the relative humidity of the outside air is relatively low.

Storage engineers can design systems controlled either manually or automatically which operate ventilating fans according to the temperature and relative humidity of the outside air. Ventilation also enables a gentle air flow to be passed through bulk lots of stored seed as and when required, thus ensuring that hot spots do not develop which would endanger the stored material.

Another form of localized heating sometimes coupled with moisture migration can occur as a direct result of convection currents. These are most likely to occur in unattended stores with relatively poor insulation and are caused by drier warm air moving from a warm spot to a cooler one within the store. On cooling, moisture is condensed which is subsequently absorbed by the drier seeds.

Refrigeration

The use of refrigeration in controlling seed store temperatures is generally confined to long-term storage of high value material, for example germplasm collections and breeders' stocks. However, refrigeration is also useful in the tropics for other categories of seed stocks.

Extra care and attention must be given to thermal insulation and structure of the store when refrigeration is to be included in the control systems.

There are four sources of heat within a seed store which the refrigeration has to cope with; these are derived from leakage from outside, despite insulation, field heat of seed and associated materials, respiration of seed in the store and incidental heat derived from lighting, other equipment, workers and external heat which enters when doors are opened.

The cooling coils of a refrigeration unit should be situated in the storage area but the compressor must be sited so that its heat is given off to the exterior of the storage room. The relative humidity of the air is reduced during the refrigeration process. The moisture condenses on the cooling coils which have to be defrosted at intervals. Although this reduction in relative humidity is an advantage, in practice the store's relative humidity is inversely proportional to its temperature and at temperatures below about 13.0°C the relative humidity is too high for safe seed storage. The following figures illustrate this point:

Temperature (°C)	Minimum relative humidity (%)
32.2	30
27.2	35
23.3	40
21.1	45
19.4	50
16.7	60
13.9	70

Dehumidification

An alternative system to refrigeration for the dehumidification of seed stores is the use of a suitable chemical desiccant in a dehumidifier.

There are two types of chemical dehumidifiers generally used in seed stores, the bed and the revolving drum. In each system the apparatus will add to the interior heat load if not carefully sited and it is therefore important that dehumidifiers be placed in the structure so that their heat is given off to the outside of the store. Silica gel, which can absorb up to 40 per cent of its own dry weight of water, is usually used for seed store systems. In the bed system the silica gel is heat dried at about 175 °C to drive off all the absorbed moisture. After cooling, air from the storage area is blown through the dried silica gel bed. When the silica gel is again in moisture equilibrium with the air it is reheated to dry it before further re-use. Some bed systems use two beds per unit, in which case one bed is dehumidifying the store's atmosphere while the other is being dried to re-activate it. The operation of alternating beds is normally controlled by a timeclock.

The revolving drum system has a desiccant bed which is divided between two airstreams. Different sections are dried or used for absorption as the bed rotates. The revolving drum systems are capable of removing more moisture from a given air flow than the bed systems.

The choice of system will depend very much on local conditions and storage requirements, so in all cases qualified and experienced environmental control engineers should be consulted.

STORAGE IN VAPOUR-PROOF CONTAINERS

There have been major developments in recent years which have led to the storage of relatively small lots (*c.* 1 kg) of vegetable seed in separate sealed moisture-proof containers.

Most of the original research and development work with this technique has been done with seed of vegetable species because of the relatively small seed-lots required by individual farmers and, with the exception of larger seeded vegetables in Leguminosae, the proportionately high value of vegetable seed per unit volume compared with cereals and other crop groups.

The principle is that seed-lots are dried to a moisture level slightly lower than they would be prior to normal open storage, and are then sealed in metal cans, packets or other suitable moisture-proof containers. As a result of this containerization or packaging, each seed-lot is in its own environment and may be stored at ambient temperature and relative humidity for one or two years, or even longer, with little or no deleterious effect on germination. In fact, provided that careful attention has been given to important details such as drying, and the determination of moisture content, and appreciating that different seed-lots can differ in their composition, it is possible that the potential storage life of a seed-lot at 8 per cent moisture content can be doubled if 1.0 per cent more moisture is removed before sealing.

Seed moisture before vapour-proof packaging

It is of fundamental importance that seed moisture is reduced to a satisfactory and safe level before the seed is sealed in vapour or moisture-proof containers. This is generally 2–3 per cent lower than for other forms of storage or packaging in non-moisture-proof containers. The reason for this is that the atmosphere within a moisture or vapour-proof container will equilibrate to the

Table 5.3   Satisfactory moisture levels for seed to be stored in vapour proof containers or packages (after James, 1967)

Family	Vegetable	Maximum per cent seed moisture
Gramineae	Sweet corn	8.0
Alliaceae	Onion, leek, chive, Welsh onion	6.5
Chenopodiaceae	Beet, chard	7.5
	Spinach	8.0
Cruciferae	Cabbage, broccoli, cauliflower, collards, Chinese cabbage, kale, turnip, rutabaga, kohlrabi, Brussels sprouts, mustard, radish	5.0
Leguminosae	Snap bean, lima bean, pea	7.0
Umbelliferae	Carrot, celery, celeriac	7.0
	Parsnip	6.0
	Parsley	6.5
Solanaceae	Tomato	5.5
	Pepper	4.5
	Eggplant	6.0
Cucurbitaceae	Cucumber, musk-melon, squash, pumpkin	6.0
	Watermelon	6.5
Compositae	Lettuce	5.5

moisture level which is in the seeds. This results in a long-term relative humidity which is too high for safe storage. For example, if sweet corn seed with a 13 per cent moisture content is sealed in a moisture-proof container the enclosed storage atmosphere will equilibrate at a relative humidity of approximately 65 per cent, which is too high for safe seed storage. In addition some storage pathogens will become active within the sealed container and the seeds respiration rate will be relatively high.

The majority of species grown as vegetables have seed with a safe moisture content during sealed storage of between 8 and 9 per cent, but in some species it is less than 8 per cent (Table 5.3).

The storage and distribution of vegetable seed with low moisture levels in vapour-proof containers has been widely adopted throughout the world. There is a range of containers used for this purpose which are discussed in the next section. The most important use of vapour-proof containers or packages is probably the shipping and marketing of high value seed, such as hybrid tomatoes to the humid tropics, but these vapour-proof containers have a major influence on seed quality control in temperate regions as well.

Some seed companies in temperate regions store their processed vegetable seed in a section of their warehouse which has a controlled atmosphere, and the operation of filling and sealing moisture-proof containers is done within the store. In arid areas seed producers containerize the seed from open store, usually immediately after drying, but first checking the moisture content of each seed-lot and further drying it if necessary.

SEED STORE MANAGEMENT

The level of hygiene within the store will have a long-term effect on seed quality and longevity. Only seed which has been through the final stages of processing should be taken into the store. All other materials should be excluded. In practice it is sometimes tempting to use seed stores for short-term holding-over of other plant materials such as dormant bulbs or selected fruit waiting for seed extraction, but this misuse of the seed store can lead to the introduction of storage pests and pathogens. Plant materials other than seeds are very likely to add to the moisture content of the storage atmosphere. Other sources of water and moisture should also be excluded in order to discourage rats.

A comprehensive programme for rodent prevention should be organized from the outset, rather than waiting for control measures to become necessary later. The possibility of rodent infestation will depend on the method of containerization within the store as well as on the location and local conditions.

Rodent prevention and control programmes include the use of rodenticides such as the blood anti-coagulants or other proprietary poisons. The material should be used in accordance with the manufacturer's instructions and current legislation relating to use of poisons and safety. Recent developments to deter rodents from stores and warehouses include the installation of sonic sound systems which are undetectable by workers. The repetition of sound recordings of the calls of birds of prey are also being used.

Seed stores should not be used for storage or shelter of machinery, apparatus or any other materials not directly involved with the stored seed. Additional apparatus and materials make it difficult or even impossible to maintain a high degree of cleanliness. All surfaces should be kept clean and ideally floors should be cleaned with vacuum cleaners in preference to brooms in order to minimize build-up dust. Any waste materials should be removed from the store as soon as they are accumulated and disposed of by burning as far away from the store as practical.

Seed stores have a relatively high fire risk due to the dry nature of seed and the possibility of dust in the air. It is therefore vital that adequate fire prevention measures are formulated and all staff made acquainted with them.

A system for entering, locating and retrieving seed-lots should be adopted. The system must take into account the need for sufficient space between seed-lots for access and air circulation. Small seed-lots should be on suitable shelving and large quantities in bags or sacks should be neatly stacked on pallets.

The most sophisticated seed stores belonging to the larger seed companies have a computerized and fully automated retrieval system. But the main criteria for any system, regardless of size of operation or level of sophistication, is that all bags, cans, etc. must be clearly labelled on both the inside and outside. Labels must be firmly attached; adhesive labels on tins should not peel off when subjected to the storage environment. The labels should be written in accordance with the inventory system adopted which should maintain a record of each seed-lot's year and other details of origin, designation, samples tested, quantities removed and balance remaining. This information will then always tally with stock books.

The structure and fabric of the buildings should be regularly inspected by competent staff. Any deterioration or damage must be restored immediately by competent tradesmen.

REFERENCES

Delouche, J. C., Matthes, R. K., Dougherty, G. M. and Boyd, A. H. (1973) Storage of seed in sub-tropical and tropical regions, *Seed Sci. Technol.* **1**, 671–700.

Harrington, J. F. (1959) In *Proceedings 1959 Short Course for Seedsmen*, pp. 89–107. Mississippi State University.

Harrington, J. F. (1972) Seed storage and longevity. In *Seed Biology* Volume III, T. T. Kozlowski (ed.), pp. 145–245, Academic Press, New York and London.

James, E. (1967) Preservation of seed stocks, *Advances in Agronomy* **19**, 87–106.

FURTHER READING

Dossat, R. J. (1978) *Principles of Refrigeration*. John Wiley, New York and London.

Ellis, R. H. and **Roberts, E. H.** (1981) The quantification of ageing and survival in orthodox seeds. *Seed Sci. Technol.*, **9**, 373–409.

ISTA (1973) Seed storage and drying, *Seed Sci. Technol.*, **1**(3).

Justice, O. L. and Bass, L. N. (1978) *Principle and Practices of Seed Storage*. USDA Handbook No.506. Castle House Publications, Tunbridge Wells, and USDA, Washington.

Mackay, D. B. and Flood, R. J. (1970) Investigations into crop seed longevity 4. The viability of brassica seed stored in permeable and impermeable containers, *J. Nat. Inst. Agric. Bot.* **12**, 84–99.

Roberts, E. H. (ed.) (1972) *Viability of Seeds*. Chapman and Hall, London.

Roberts, E. H. (1975) The problems of long-term storage of seed and pollen for genetic resources conservation. In *Crop Genetic Resources for Today and Tomorrow*, O. H. Frankel and J. G. Hawkes (eds). University Press.

Roos, E. E. (1980) Physiological, biochemical and genetic changes in seed quality during storage, *Hortscience*. **15(6)**, 781–4.

6 SEED HANDLING, QUALITY CONTROL AND DISTRIBUTION

SEED COMPANY RECORDS

These record the successive stages from field production to assembly in the seed company's premises. In some countries it is a legal requirement to maintain records of the origin and identity of each seed-lot. Well kept records are also essential for future planning at farm level and for dealing with problems relating to seed quality, such as growing on tests, admixture or cultivar purity (i.e. genetical quality), and seed-borne pathogens.

Records during planning and production

These should include the following information: rotation and cropping frequency; applications of pesticides throughout the crop (including pre-sowing or pre-planting) and partial sterilization and soil fumigants; cultural details e.g. sowing rates and dates of sowing, planting and other operations, applications of fertilizers during preparation and growth of the crop; roguing stages and dates with identity of responsible staff, details of off-types and percentage removed or selected plants retained; observations on effective isolation and seed yield.

Records of a seed-lot

These include a stock or lot number which is designated on arrival of the seed-lot and against which all subsequent information on that lot is recorded. Stock numbers are allocated from a stock book and may be formed from a coded system of digits and letters including stock number and information such as year of multiplication, seed generation, standard of processing, and processor's identity.

All seed-lots received are entered into the stock book including seed-lots produced by the company, contractors of other seed companies. The records include year of multiplication, pre-cleaning and processing operations and quantity after processing. Additional information recorded includes weight of a sample (10 g is normally an appropriate quantity for vegetable seeds) and the results of germination, purity and growing on tests. Further data such as results of annual generation tests are added during the time that the seed-lot is kept by the company including seed treatments during storage and prior to dispatch.

SECURITY OF SEED STOCKS

Factors affecting longevity of stored seeds were discussed in Chapter 5 but the value of seed stocks is again emphasized here. Particular attention must be given to security, fire and vermin or insect attack, and to ensuring that the identity of seed-lots is not lost or mistaken. It is customary to have duplicate labels on both the outside and inside of containers, and staff must be trained to check both of these when handling the seed-lot. It is advisable that only authorized members of staff be allowed to take from or add to seed stores.

MONITORING SEED QUALITY

The reasons for assessing the quality of a seed-lot are numerous and include determining quality, value, purity and moisture content. The different aspects of quality, which include potential germination, vigour, purity, health and trueness to type, are used to evaluate the sowing value of a seed-lot, i.e. the true value to farmers and growers. They may be used also to determine the commercial value, i.e. the price paid to the contractor who produced the seeds or vendor in a subsequent commercial transaction, and ultimately to protect the final purchaser of the seed who uses it to produce a vegetable crop. Different evaluations may be done in the pathway of the seeds from original producer through to the grower who sows them but each evaluation generally follows an internationally agreed and accepted procedure.

The role of the International Seed Testing Association (ISTA)

ISTA publishes methods of seed evaluation which are accepted internationally by 59 subscribing countries which in turn accredit 137 official seed testing stations and 173 personal members to the Association. The present distribution of member countries is Africa 9, Asia 10, Australasia 2, Europe 26, Middle East 4, Latin America 6 and North America 2. The Association publishes standard procedures of related topics such as seed testing, and promotes their adoption by the seed trade up to international level. In addition, ISTA technical committees promote and co-ordinate research into different aspects of seed technology, and hold a tri-annual congress which includes meetings of its technical committee, reviews of its rules for seed testing and a symposium on topics related to seed technology. ISTA publishes *Seed Science and Technology*; other examples of ISTA publications are listed under 'References' and 'Further Reading' at the end of this chapter.

In North America the Association of Official Seed Analysts plays a similar role to ISTA and in practice the two associations have a strong affinity and mutual respect.

The international analysis certificate

Seed testing laboratories which have been approved by the Executive committee of ISTA and which use tests in accordance with the rules are authorized to issue the special international seed analysis certificate which is of high repute in the international seed trade and which assists the movement of seed-lots between different countries.

THE ORGANIZATION OF SEED TESTING

The general aims of legislation relating to seed quality are to ensure that seeds offered for sale conform to minimum required standards of quality, viz. cultivar, health, vigorous germination and freedom from adulterating material including weed or other crop seeds. Coincidental to these objectives are protection of the vegetable grower from unscrupulous vendors. In countries where the seed industry is relatively undeveloped the latter point may be more important but once a seed industry is established, it is the monitoring of the different facets of seed quality that gains in importance, especially when certification schemes are introduced.

Seed testing stations

In order to monitor seed quality it is necessary to have impartial seed testing stations. Such stations are financed by governments and usually work within the framework of ministries of agriculture or as part of an institute set up and financially aided by the ministry of agriculture to oversee advice on seed quality. In the UK this is done for England and Wales by the Official Seed Testing Station (OSTS) within the framework of the National Institute of Agricultural Botany (NIAB) based at Cambridge. The OSTS at East Craigs, Edinburgh, covers Scotland, and the OSTS at Crossnacreevy, Belfast, deals with Northern Ireland. Close liaison between the senior officers of the stations ensures that their very high standards are maintained and procedures standardized.

In addition to these three main stations the Ministry of Agriculture grants licences to approved seed laboratories within commercial companies. Officers of the OSTS are responsible for their technical supervision, staff training for approved qualifications, and advising on related problems. Licensed seed testing laboratories are also referred to as satellite stations. In addition to pro-

viding information on seed quality, such as germination and purity, required
to satisfy the legal aspects of sales, they provide an in-house service, e.g.
moisture determinations, monitoring the level of purity between processing
operations and doing routine germination, vigour and health tests while the
seeds are in store.

Seed testing

The sampling and range of tests to assess seed germination, vigour, purity,
health and moisture content are all done by standard methods approved by
ISTA. Genetical quality (or cultivar purity) is evaluated by a growing-on test
in field plots. Descriptions of the techniques used in seed testing are outside
the scope of this book and the reader is referred to the 'Further Reading' list at
the end of this chapter.

Packaging

A wide range of materials and types of containers is used for seeds. The types
of materials and containers suitable for vapour-proof storage were discussed
in Chapter 5. Some seed stocks are kept for relatively long periods in controlled storage (see Chapter 5) before placing them in vapour-proof containers. Other seeds, especially the relatively bulky species in Leguminosae,
are distributed by seed companies without vapour-proof packaging.

The range of materials used for containers and packages includes burlap
(a coarse canvas), cotton, paper (bleached sulphate and bleached Kraft papers
are both used), polythene, hessian (usually known as gunny bags), aluminium
foil, tinfoil and tin-plate.

Some modern packets and containers are made from laminations of two
or more materials, such as aluminium foil and polythene; aluminium foil,
paper and polythene or laminations of paper and asphalt. Glass containers are
used frequently for reference collections for seed identification (Fig. 6.1) but
are not suitable for seed distribution.

A selection of seed containers is shown in Fig. 6.2.

The choice of material and size of container depends to some extent on the
level of development of the seed industry, packaging machines and the destination, type of market, mode of distribution, protection required from hazards such as rain, high relative humidity, rodents, insects, pathogens and the
amount of handling during transit.

Size of containers

The quantity of seeds and size of container of packet are usually adjusted
according to the intended market, for instance relatively small quantities in

Fig. 6.1   A reference collection of seed samples.

Fig. 6.2   Different types of seed containers.

packets for the amateur or private gardeners. The exact quantity may be adjusted to potential germination, purity and the selling price per packet. For professional growers the quantity is usually in units of a gram or kilogram. A notable exception to this is the sale of hybrid sweet corn seeds for the North American market, which are sometimes packaged in quantities suitable for sowing a specified unit area. The content of vegetable seed packages is rarely based on a specified (or pre-determined) number of seeds although this is frequently done for the flower seed trade. Possible exceptions are high value, relatively large seeds such as F_1 hybrid cucumber seeds for the amateur market.

Modern seed companies have very sophisticated packaging lines in other warehouses which are capable of automatically filling containers by machines, delivering pre-determined quantities of seeds (Fig. 6.3) and applying labels to the packets.

Labelling

The label on a seed package is the normal way of immediately identifying the contents. Labelling systems used in seed stores and for warehousing usually conform to the system devised for the company's stock book.

Labelling for the market usually has to meet the legal requirements for the country in which the seeds are to be marketed and includes information conforming to certification schemes. Seed companies also provide additional information according to the market outlet (i.e. private or commercial grower). The following information may therefore be put on labels:

Name of the seed company and its trade-mark.
Species and cultivar name.
Year of packaging.
Germination, purity and statement regarding noxious weed seeds.
Seed treatments (e.g. Thiram treatment of celery seeds).
Batch number (this will give the seed company immediate
   reference to the seed-lot in any subsequent enquiry or complaint).
Class of seed (internationally agreed colour codes are used for different
   seed classes).

Packages for the amateur or private market usually have a pictorial label which gives an indication of the species and cultivar's characters. In addition basic cultural information may be added such as sowing date, husbandry methods, including frequency of successional sowings. Where seed companies export to several different countries some of this information is presented diagrammatically or in several languages.

PROMOTION

A seed company's promotion of its products, and the activities of governments and other agencies in developing countries to focus farmers' and grow-

Fig. 6.3   Mechanized filling, sealing and labelling of seed packets.

ers' attention on improved seeds, are important aspects of marketing and adoption of new cultivars.

In addition to their commercial role, these activities play an important part in agricultural and horticultural development. Other influences or sources of information on the use of improved seeds or new cultivars are the advisory and extension services, recommended lists and the press.

Catalogues

The majority of seed companies produce an annual catalogue and price list, normally in sufficient time for growers to place their seed orders before the sowing season commences. Catalogues are usually produced for specific consumer groups, i.e. amateur or commercial growers. Many companies list flower and vegetable seeds in the same catalogue for the amateur market. Cultivars are listed for each crop and it is customary for a popular description of the cultivar to be included with particular reference to specific season of use, morphological type and resistance to specific pathogens. Additional information may include special recommendations, e.g. success of the cultivar in trials. The catalogue number of each item is usually used when placing orders and conforms to reference numbers used in the dispatching warehouse. Seedsmen's catalogues usually contain other relevant information including conversion tables, numbers of seed per unit weight, sowing rates, programmes for successional production and may offer optional services for some species such as pre-sowing treatments for specific pathogens. Other items included are statements relating to the current seed laws and trading conditions.

Demonstrations, open days and shows

There is probably no better way of illustrating the merits of specific seed stocks or cultivars to farmers and growers than growing them on demonstration plots under the same ecological conditions as prevail in the vegetable production industry. Seedsmen therefore organize demonstrations of crops grown from their material. At appropriate times of the year 'farm walks', 'open days' or 'cultivar demonstrations' are organized. These may be held in conjunction with other interested organizations such as advisory services, processing organizations or growers' co-operatives. In developing countries demonstrations may be organized at village level by organizations involved with population health and nutrition, aid programmes and education.

Public relations

The public relations activities of seed companies often include a technical advisory service, and technical representatives who, in addition to their sales activities (see below), are also involved in offering advice and dealing with complaints. They generally have a very good local knowledge of the industry. Reputable sections of the seed industry are often too hastily blamed for poor seedling emergence or crop stand and the diplomatic attitude of a seed company's technical staff can sometimes be backed by results of laboratory or growing-on tests. Conversely, a problem occurring with the same original seed-lot distributed to different growers may be resolved by reference to test records. There has been a general increase in the amount of printed technical information available from seed companies especially that produced for commercial vegetable growers.

The participation of seed companies at shows and exhibitions is also a way of displaying cultivars and merits of a particular company. Some seed companies sponsor prize money at local shows for winners who have used their seeds. There has been a general decline in seed companies' activities in UK shows, especially at county level and below, due to the very high cost of producing and mounting exhibits. However the advent of permanent show grounds has assisted institutions which support the vegetable industry (e.g. ADAS [Agricultural Development and Advisory Service] and NIAB in the UK) to organize demonstrations involving the growing crop. The concept of 'grower education' in a modern vegetable production industry involves a multi-organizational approach including educational, research, extension and commercial activities and the closer the liaison between these different types of organization the better the prospects for the grower and the industry as a whole.

PRICE STRUCTURE

Marketing of demand factors

The demand for seeds in countries with an established vegetable production industry is relatively stable compared with that in developing countries in which there may be a rapid transition from total reliance on farmers' own saved seeds to seeds produced by government agencies or commercial organizations. The relatively complex operations in vegetable seed production, involving factors such as biennial species, vernalization requirements, utilization of F_1 hybrids and skill required to maintain genetical quality, tend to minimize growers' motivation to produce their own seeds.

There is evidence that in countries where new seed industries have been set up it is beneficial in the long term to sell the seeds at a price level to cover all the production costs plus a profit for the vendor (Douglas 1980). Generally, the more realistic the charge made for seeds produced by government agencies in the early stages of seed industry development, the sooner the private sector will wish to participate. There is therefore a better basis for a realistic seed price in the vegetable seed industry than perhaps in other areas of agriculture, especially in countries with a recently created demand for increased quantities of seeds. However, some government initiatives relating to legislative requirements (e.g. seed testing and certification schemes) will tend to increase the overall costs of production.

The components of seed price can be broadly classified into two groups, viz. direct costs and indirect costs.

Direct costs

These include the factors which are directly related to the quantity of seeds marketed, such as costs of stock, processing, containers and transportation.

Indirect costs

These are the factors which are not directly related to the quantity of seeds marketed and include activities relating to promotion such as maintenance of trial grounds, production and distribution of the catalogue.

DISTRIBUTION

The relatively high value and perishable quality of vegetable seeds makes the need for care during transportation and distribution a high priority. The deterioration of seed quality is generally minimized by the use of vapour-proof containers and packages (discussed in Chapter 5). However, factors which may contribute to deterioration of the containers, labelling system and seeds not transported in vapour-proof modules include changes in temperature and relative humidity, condensation on arrival in high temperature zones as a result of chilling during airfreight, contamination by other items of cargo during short-term storage and distribution. It is therefore extremely important that the seed company and its agents appreciate the hazards to which the product is exposed during marketing.

A commercial company marketing seeds internationally usually has a marketing manager for each region. Within a region there are distributive agents or the company may have its own direct sales outlets.

On a national basis companies may sell via agents, wholesale or retail outlets, or their own sales staff whose role it is to promote their company's products in a specified area and to obtain orders. In this latter case their sales staff may also be responsible for the supply of stock to outlets such as garden centres. In addition most seed companies sell direct to growers.

REFERENCE

Douglas, J. E. (1980) *Successful Seed Programs, a Planning and Management Guide.* Westview Press, Boulder, Colorado.

FURTHER READING

Bekendam, J. and Grob, R. (1979) *Handbook for Seedling Evaluation.* ISTA, Zurich.
Gregg, B. R. (1983) Seed marketing in the tropics, *Seed Sci. Technol.* **11**, 129–48.
Gregg, B. R., Delouche, J. C. and Bunch, H. D. (1980) Inter-relationship of the essential activities of a stable, efficient seed industry, *Seed Sci. Technol.* **8**, 207–27.
Heydecker, W. (1972) Vigour. In *Viability of Seeds*, E. H. Roberts (ed.), pp. 209–52. Chapman and Hall, London.
ISTA (1964) Variety purity examination, *Proc. Int. Seed Test. Ass.* **29**(4). ISTA, Zurich.
ISTA (1976) International rules for seed testing, *Seed Sci. Technol.* **4**, 3–49.
ISTA (1976) International rules for seed testing, Annexes 1976, *Seed Sci. Technol.* **4**, 51–177.
ISTA (1981) *Handbook on Seed Health Testing,* Section 2. ISTA, Zurich.
Justice, O. L. (1972) Essentials of seed testing. In *Seed Biology*, T. T. Kozlowski (ed.), pp. 301–70. Academic Press, New York and London.
MacKay, D. B. (1972) The measurement of viability. In *Viability of Seeds*, E. H. Roberts (ed.), pp. 172–208. Chapman and Hall, London.
MacKay, D. B. (1978) The laboratory germination test, *Acta Horticulturae* **83**, 97–102.
Mathur, S. B. (1983) Testing seeds of tropical species for seed-borne diseases, *Seed Sci. Technol.* **11**, 113–28.
Perry, D. A. (1978) Problems of the development and application of vigour tests to vegetable seeds, *Acta Horticulturae* **83**, 141–6.
Perry, D. A. (1981) *Handbook of Vigour Test Methods.* ISTA, Zurich.
Phatak, H. C. (1974) Seed-borne plant viruses – identification and diagnosis in seed health testing, *Seed Sci. Technol.* **2**, 3–155.
Richardson, M. J. (1979) *An annotated list of seed-borne diseases*, Commonwealth Mycological Institute, Kew, England and International Seed Testing Association, Zurich, Switzerland.
Richardson, M. J. (1981) Supplement 1 to *An annotated list of seed-borne diseases* Commonwealth Mycological Institute, Kew, England and International seed Testing Association, Zurich, Switzerland.
Sargent, M. J. (1973) *Economics in Horticulture.* Macmillan, London.
Thomson, J. R. (1979) *An Introduction to Seed Technology*, Chapters 11, 14, 15, 16 and 17. Leonard Hill, London.
Ulvinen, O., Voss, Å, Backgaard, P. E. and Terning, P. E. (1973). *Testing for Genuineness of Cultivar.* ISTA, Zurich.

7 CHENOPODIACEAE

The main genera in this family which are cultivated as vegetables are:

Beta vulgaris L. subsp. *esculenta*:   beetroot, red beet
Beta vulgaris L. subsp. *cycla*:   spinach beet, chard or Swiss chard
Spinacea oleracea L.:   spinach

There are also a few other vegetable crops in this botanical group which are of local importance only. These include *Chenopodium quinoa* Willd. (quinoa), cultivated as a staple food in South America, particularly in the Andes.

BEETROOT: *Beta vulgaris* L. subsp. *esculenta*

The cultivated beetroot with its swollen root is derived from *B. vulgaris* subsp. *maritima* (L.) Thell. Sugar beet, which is an important industrial crop, and mangold (or mangel-wurzel), grown in temperate regions as a root crop for stock feed, are also forms of *B. vulgaris*; they are both closely related to beetroot to the extent that they freely inter-pollinate.

Beetroot, which is generally grown for the production of its swollen roots, is a popular crop in the Middle East, Europe and North America. Processors have shown an increased interest in this crop in recent years. It is not widely cultivated in the humid tropics, but is increasing in popularity in the cooler tropics.

Criteria for development and selection of the modern cultivars have been based on root shape and colour, and include globe-, cylindrical- and long-rooted types.

Cultivar description of beetroot

Season	suitability for early sowing resistance to premature bolting
Seed (i.e. fruit) type	multigerm or monogerm
Leaf	degree of anthocyanin approximate number of leaves to crown

	lamina, relative size, shape and degree of blistering
	colour and lustre
	character of lobes
	petiole, relative length, thickness and colour
Root	shape: e.g. flat, globe, cylindrical, conical or long (Fig. 7.1)
	crown of root: coarse or fine
	colour and intensity of colour: red or yellow
	interior quality (freedom from 'rings')

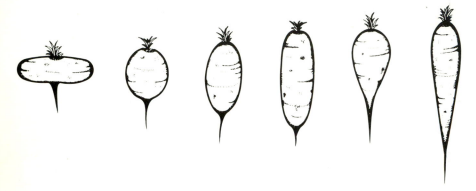

Fig. 7.1   The main beetroot shapes; left to right; flat, globe, oval, cylindrical, conical and long.

The 'seed' of beetroot is formed by an aggregation of flowers fused by their swollen perianths to form a multigerm fruit, which is a cluster of two to five seeds. Plant breeders working with sugar beet have developed monogerm types in which the flowers remain unattached to each other and the fruits therefore bear only a single seed. This character has important agronomic advantages, allowing the seeds to be sown singly for precision drilling and optimum spacing of single plants, obviating the need for singling. The monogerm character has also been transferred to beetroot, but only a few cultivars are available with this character, the majority still offered by the seed trade being multigerm.

Agronomy

Soils and nutrition

Beetroot are slightly tolerant of acid conditions and soils with a pH between 6.0 and 6.8 are suitable.

The general ratio of fertilizers applied during seed bed preparations is N : P : K 2 : 1 : 2, although some growers prefer a lower proportion of nitrogen than this and apply 1 : 1 : 2 during preparations with further nitrogen given as top dressings in the first season. A spring dressing of nitrogen is given to transplanted stecklings at the rate of 50 kg/ha. Supplementary nitrogen is also applied as a top dressing before flowering in areas with a relatively high leaching rate.

Soils with a low boron status should either be avoided or a supplementary dressing of boron applied during preparations. Ideally, a boronated fertilizer is used for the main base dressing or, if not available, boron is applied as sodium tetraborate (borax) at the rate of 1.5 to 2 kg/ha. The main symptoms of boron deficiency are black cankerous areas on the root exterior and also between the concentric rings. This interior symptom is clearly seen when affected roots are cut transversely.

Manganese deficiency occurs in some specialist production areas, e.g. in Washington State, USA, where the problem is counteracted by dressings of up to 100 kg/ha of manganese sulphate during seed bed preparation.

The main beetroot seed production areas grow the crop on the flat but in the Middle East, where furrow irrigation systems are used, the crop is grown on ridges.

Seed production methods

There are two basic methods of seed production for beetroot; these are 'seed to seed' and 'root to seed', which are defined in Chapter 3. The 'seed to seed' system is normally used only for the final stage of seed multiplication while the 'root to seed' system is used for the production of basic seed and also preferred in some areas for the final multiplication. This latter system allows for inspection and roguing of roots.

Root to seed

This system is in two stages. The first stage is very similar to the production of beet for market. Seed is sown in the mid-summer at a rate of 10–15 kg/ha in rows 40–50 cm apart. Seed is sown during July–August in areas such as California, where growing conditions are satisfactory into the autumn. The seedlings are singled as soon as possible following emergence, to allow the root to develop its characteristic shape, but this operation is not normally necessary if monogerm or rubbed seed has been precision drilled.

The roots are lifted in the autumn, selected for root characters and the desirable material stored. The lifting operation is completed before the onset of damaging frosts, and thus the timing depends on the local climate. Whatever lifting procedure is adopted, every care must be taken to avoid mechanical damage to the roots. The lifting and subsequent root handling is easier if the plants are 'topped' first by passing over the crop with a cutter bar or mower. The crown of the root must not be damaged during topping or lifting.

Root storage
The two basic systems for storage of roots are either the use of suitable buildings or storage in the field in clamps or pits. Most beetroot seed producers have developed suitable shed systems, but field storage is still used in some areas.

Storage in buildings.   The main advantage of storage in buildings is that the air temperature can usually be controlled when necessary, thus avoiding over-heating and frost damage. The optimum temperature for beetroot storage is 4–5 °C. The optimum relative humidity is between 80 and 90 per cent, although very few seed producers have sufficient facilities to control the RH. A stacked tray or crate system is very suitable and can be co-ordinated with field operations at lifting and planting times. Where possible the boxes of selected roots are air dried under cover before stacking in the store.

Field storage.   There are several versions of field storage, which include clamps and pits. The roots are arranged in pyramids or ridges on well drained sites. In both of these versions the roots are stacked from 60 to 200 cm above ground level. The piles of roots are covered with straw, which is held in position by loose soil. Polythene sheets can also be used to exclude the rain and protect from frost, but care has to be taken to avoid condensation. Straw funnels or chimneys in the ridges reduce the risk of condensation.

The roots are re-planted in the spring, as early as local conditions allow, in rows 100 cm apart, with approximately 30 cm between roots within the rows. The roots must be set upright with their crowns at final soil level. Some growers plant into a relatively loose soil and pass a roller over the field after planting in order to firm the soil around the roots. This reduces drying out and is claimed to assist the early establishment of new fibrous roots.

In countries with a very well organized and specialized vegetable seed industry, the roots are produced and stored by the seed companies and supplied to growers who plant them and produce the seed on a contract basis.

Seed to seed

In this system a later sowing date is adopted compared with the root to seed system described above and the plants remain in the field all winter, so the method is not suitable in areas with severe frosts. The actual sowing date depends on local climates, but August-September is the main time. The sowing rate is approximately 12 kg/ha which produces sufficient stecklings to plant up 4 ha. The seed is sown in a four row bed system with 25–30 cm between the rows, with a bed width of 110 cm. Sowing rates are adjusted according to seed type to give an optimum plant density in the beds of 200/m².

The stecklings are transplanted in the early spring. But in some of the Mediterranean areas spring transplanting is difficult because the soil is very wet, and late autumn transplanting is adopted.

The optimum size of transplant is *c.* 2.5–2.75 cm (weighing *c.* 40–45 g). Although the trimming of transplants facilitates the operation, especially the long rooted cultivars, the swollen tap root must not be cut.

The planting distances are 60 cm between rows and 45–60 cm within the rows. In drier areas the transplants are frequently irrigated until established.

Flowering

The beetroot is a quantitative long-day biennial with a cold requirement for flower initiation. The detailed investigations into this requirement were first made by Chroboczek (1934). More recent work has demonstrated that exposure of the ripening seeds to low temperatures can reduce the subsequent low temperature requirement. This early vernalization sometimes contributes to the incidence of bolting in the first year.

The inflorescence emerges from the growing point relatively early in the spring. Some seed producers 'top' this flowering shoot when it is *c.* 40–50 cm high. It is claimed that this increases seed yield by reducing the duration of flowering and concentrates the seed maturity period, which in turn reduces losses from shattering.

Pollination

Beetroot flowers are predominantly wind pollinated although some insect pollination by Diptera occurs.

Isolation

It is generally accepted that pollen of *Beta vulgaris* is wind-borne over relatively long distances and sufficient isolation should therefore be ensured. Most authorities stipulate isolation distances of at least 500 m between cultivars of the same type (e.g. red globe) and at least 1000 m between different types of cultivar (e.g. between red globe and cylindrical types).

Beetroot is cross-compatible with the other subspecies of *Beta vulgaris* (i.e. sugar beet, mangolds, spinach beet and Swiss chard) and adequate isolation of different seed crops has to be ensured. This is usually accomplished by a zoning scheme (discussed in Chapter 3). In the UK scheme (Anon. 1981), the stipulated minimum isolation distance between the different types of *Beta vulgaris* is 1 km, although the recommended distance in some states of the USA is at least three times this.

When seed of high genetical quality is required or pollen contamination is suspected. the discard strip technique (Dark 1971) can be used (details in Chapter 3). Although this technique was developed for sugar beet seed production, the same principles apply with other wind-pollinated *Beta vulgaris* types.

Roguing stages

The roguing of plants for beetroot seed production is considerably more thorough when the root to seed system is used. The seed to seed system does not allow the mature root characters to be observed.

Seed to seed
The main roguing is done at lifting and re-planting, although plants which bolt prematurely can be removed before lifting.

1. Lifting   Discard plants showing any of the following: incorrect leaf shape and colour, premature bolting, incorrect root shape, seed-borne pathogens.

2. Re-planting   Characters as described above.

Root to seed

1. Before cutting tops for lifting   Remove plants showing any of the following: incorrect leaf colour and morphology, early bolters and seed-borne pathogens.

2. Lifted roots   Discard roots which are not true to type. Shape, size, crown and surface corkiness should be taken into consideration. Examples of the range of root types which may occur in a red globe cultivar are shown in Fig. 7.2.

Fig. 7.2   Range of root types observed in a red globe beetroot cultivar.

3. Re-planting   If roguing has been done in accordance with stage 2 described above, no further roguing is required, although roots showing storage diseases should be discarded.

4. Bolting plants (before 'topping')   Remove plants showing incorrect leaf shape, leaf colour, vigour and seed-borne pathogens.

Improvement of basic stock

The production of basic stock seed is always done by the root to seed method. All the crop is rogued at all the stages described above for root to seed production, and in addition interior root characters are observed before re-planting by cutting a thin wedge of flesh. This enables the degree of colour differentiation between the vascular and parenchyma tissues to be observed (Fig. 7.3).

Fig. 7.3 Beetroots with wedges cut out to examine internal root quality.

Alternatively a cork borer is used to remove a sample of root flesh. Unsatisfactory roots are discarded before the remainder are planted. Some seed producers replace the wedge or core while others prefer to dust the exposed surfaces with a fungicide.

In addition to this inspection of interior root characters, the discard strip technique can be used for production of stock seed.

Harvesting

The harvesting of beet seed commences when the 'fruits' at the bases of inflorescent side shoots mature. By this stage the fruits have turned from green to brown. An additional check is to cut transversely a sample of ripe fruits. Unripe fruits are milky when cut and ripe fruits are mealy. Seeds ripen successively from the bases of the side shoots to the terminal point. Care is needed to determine the optimum time for cutting because the immature seeds shrivel if cut too early, and if cut too late seeds are lost as a result of shattering.

Ripening beetroot stems tend to be prostrate rather than vertical. The method of cutting depends on the scale of operation. Large-scale producers in the USA use a swather, but for small-scale production, basic seed or larger scale production with relatively low hand labour costs, the crop is cut with knives or hooks.

The cut stalks are left in windrows to dry and carefully turned once or twice. In areas where autumn rain is a problem the cut stalks are tied in bundles and dried on 'four poles' placed in shocks (stooks) or alternatively open sheds. Large heaps of cut material are placed on tarpaulins, or polythene sheets, to avoid loss of seed from shattering. The cut material can take from three to fourteen days to dry according to the air temperature and rainfall.

Threshing

After drying, the material is threshed by a stationary thresher or a combine. The dry straw of beetroot seed is extremely brittle, and it is therefore important to use a relatively low cylinder speed and air blast. Concave openings must be wide in order to avoid producing too many small pieces of straw as it is difficult to separate these off afterwards. There is relatively little chaff from beetroot material.

The final separation of beetroot 'seeds' from the small pieces of plant debris is done on a gravity separator (described in Chapter 4).

Seed yield

A satisfactory yield of beetroot seed in most areas of the world is approximately 1000 kg/ha, although up to twice this amount is achieved in the USA.

Seed weight

The 1000 grain weight of multigerm beetroot 'fruits' is *c*. 17 g; rubbed and graded 'seed' has the lower 1000 grain weight of *c*. 10 g, but as the size of the 'fruits' varies between seed-lots these figures are only given as a guide.

The main seed-borne beetroot and Swiss chard pathogens with common names of the diseases they cause

Pathogens	Common names
Alternaria alternata (Fr.) Keissler, syn. *Alternaria tenuis* Auct.	Seedling rot, leaf spot
Cercospora beticola Sacc.	Leaf spot
Colletotrichum dematium (Pers. ex Fr.) Grove f. *spinaciae* (Ell and Halst.) Arx, syn, *C. spinaciae* Ell. and Halst.	
Erysiphe betae (Vaňha) Weltzien, syn. *E. communis* (Wallr.) Fr. f. *betae* (Vaňha) Jacz.	Powdery mildew
Fusarium spp.	Blackleg
Peronospora farinosa (Fr.) Fr., syns. *P. schachtii* Fuckel, *P. effusa* (Grev.) Rabenh.	Downy mildew

Pathogens	Common names
Pleospora betae (Berl.) Nevodovsky, syns. *P. betae* Bjoerling, *P. bjoerlingii* Byford, *Phoma betae* Frank	Blackleg, damping-off, leaf spot
Ramularia beticola Fautr. and Lamb	Leaf spot
Corynebacterium betae Keyworth, Howell and Dowson	Silvering of red beet
Pseudomonas aptata (Brown and Jamieson) Stevens	Bacterial blight, leaf spot, black streak, black spot
Viruses	Arabis mosaic virus, Raspberry ringspot virus Tomato black ringspot virus (beet ringspot virus) Lychnis ringspot virus
Ditylenchus dipsaci (Kühn) Filipjev	Eelworm canker

SWISS CHARD, CHARD, SPINACH BEET: *Beta vulgaris* L. subsp. *cycla*

This crop which is also a subspecies of *B. vulgaris* has been developed and selected for its broad leaves and wide petioles. It is a biennial and is cross-compatible with the other cultivated types of *B. vulgaris* (beetroot, sugar beet and mangolds).

The crop is grown commercially in Europe (especially France and Italy) and North America. It is also a popular garden vegetable in these areas, the Mediterranean, Asia and the higher elevations of the tropics and sub-tropics where it is grown in the dry seasons.

Cultivar description of spinach beet

Season of production	suitability for specific seasons, resistance to early bolting, adverse weather conditions and low temperatures
Leaf blade	colour: red, green or yellow-green width degree of crinkling
Petiole	colour: red, green, yellow-green or bi-colour relative width and length

Agronomy

The soil, nutrient requirements and seed production methods are similar to beetroot. However, because the leaves are relatively large a lower plant population is used. Seed is drilled at the rate of 6 kg/ha in rows 90 cm apart. Seedlings are singled and thinned to a final stand of 30–45 cm within the rows.

As with the production of beetroot seed, both seed to seed and transplanting systems are used. The transplanting system allows a better opportunity for plant inspection and is the only method used for the production of basic seed. Unlike beetroot, the transplants are not stored. Commercial seed is produced in areas of the USA where the winter climate is sufficiently mild for plants to survive in the field and be transplanted in the spring. In Europe, especially Southern France and Italy, the selected plants from a summer sowing are transplanted in the autumn and overwintered in unheated plastic structures (Fig. 7.4). The plastic protection is removed in the spring.

Fig. 7.4   Selected plants of Swiss chard which have overwintered under plastic.

Isolation

The minimum isolation distance between any similar coloured cultivar is 1 km. This is doubled for different coloured cultivars. The recommendation or stipulation for Swiss chard isolation from other types of *Beta vulgaris* is frequently at least 5 km in the USA, but depends on the importance of sugar beet seed production in the area. The importance of sugar beet seed production tends to dictate the isolation distances between the different types of *Beta vulgaris* in the major seed production areas of the world.

Roguing stages

Seed to seed

With this system roguing is done before the onset of winter and again in the spring. Early bolting plants are removed in the autumn and on both occasions the leaves and petioles are inspected for trueness to type. Rogue plants and those with seed-borne pathogens are removed.

Harvesting

The stage of ripeness for harvesting is similar to beetroot. Swiss chard inflorescences are from 2–3 m high, but after fertilization tend to become more horizontal and meshed together. This is a similar situation to sugar beet and a sugar beet harvester with two cutter bars and special windrower is very appropriate for large-scale operations. For smaller scale production the crop is cut with knives or hooks (as described for beetroot) and dried in windrows.

The seeds are threshed and separated by the same processes as described for beetroot.

Seed yield

The seed yield is approximately 1.1 tonnes/ha, with good yields of up to 2 tonnes/ha reported by specialist producers in the USA.

1000 grain weight

The 1000 grain weight of Swiss chard is approximately 17 g.

Seed-borne pathogens

The seed-borne pathogens are the same as those listed for beetroot.

SPINACH: *Spinacea oleracea* L.

The common name *spinach* is given to different species of leafy vegetables in different parts of the world. Seed production of the so-called European spinach (*Spinacea oleracea*) is discussed in this section. African spinach (*Amaranthus* spp.) is dealt with in Chapter 16.

Spinacea oleracea was introduced into Europe from South–West Asia probably in the fourteenth century, and it later reached North America. It is now widely cultivated in all temperate regions. Cultivation in the tropics is not important except in the higher elevations.

The modern cultivars have been developed for their abundance of edible leaves. In the last three decades the crop has become important to processors, who market it as a canned *purée* and in frozen packs.

Spinach cultivars are mainly classified according to seed type (round or 'prickly'; Fig. 7.5) and leaf characters.

Fig. 7.5   Seeds of round and prickly spinach.

Cultivar description of spinach

Seed type	round or prickly
Season	sowing and production: relative earliness, suitability for sowing in long days, resistance to early bolting
Leaf	approximate number of leaves before bolting colour shape texture: smooth or crinkled pose
Petiole	length colour

Resistance to downy mildew (*Peronospora spinaciae* Laub.)

Agronomy

Soils and nutrition

The optimum soil pH is 6.0 to 6.8, and if the pH is below 6.0 an appropriate dressing of a liming material is given during preparations.

The general N : P : K recommendation is in the ratio of 1 : 2 : 2 applied during the final stages of seed bed preparation. Supplementary top dressings of nitrogen are applied before and after bolting. Spinach plants are prone to lodging after flowering commences, and supplementary applications of nitrogen should therefore be applied according to local experience and the amount of nitrogen lost by leaching.

Soils known to be infested with *Fusarium* or *Verticillium* wilt pathogens must not be used for seed production.

Sowing

In the USA sowings are made either in the autumn or spring. Canadian producers sow only in the autumn. In Europe sowings are made at either time according to location and hardiness of the cultivars.

Seed is sown at the rate of 6 kg/ha, in rows 45–60 cm apart. Seedlings are not normally thinned except for basic seed production.

Flowering

Spinach is a typical long-day plant. Flowering occurs in unvernalized plants but is hastened by previous chilling.

Flower type

Populations of spinach are composed of plants which are either male, female or hermaphrodite. Male plants tend to flower before female plants. There also tends to be a correlation between the dioecious plants and leaf size and number; male plants produce fewer and smaller leaves before flowering whereas female plants produce more and larger leaves before flowering. Typical male and female plants at flowering are shown in Fig. 7.6.

Pollination and isolation

Spinach is mainly wind pollinated. Recommended isolation distances are therefore up to 1000 m in some countries, although some authorities stipulate only 500 m for production of commercial seed cultivars within the same type (e.g. leaf type and seed type).

Fig. 7.6   Female (left) and male (right) spinach plants.

Spinach roguing stages

1. Before main flowering (when rosette formed), remove non–rosetting and early flowering male plants and plants not true to type.
2. When flowering has commenced, as for stage 1.

Plants which are susceptible to specific pathogens, e.g. *Peronospora spinaciae* (downy mildew) and mosaic (cucumis virus 1) are removed during roguing; also plants showing symptoms of seed–borne pathogens.

Hybrid seed production

The ratio of female to male rows is usually 6 : 2 or 14 : 2.

Specific instructions for roguing and harvesting are given by the maintenance breeder. Roguing for hybrid seed production usually includes the removal of male plants from the female rows. This is usually done twice to ensure complete removal of the males from female rows. Some hybrid lines are produced by crossing round and prickly seeded parents, and this enables the seed products of the parents to be harvested together and subsequently separated during processing (Sneep 1958).

Harvesting

In the seed producing areas with relatively calm and dry weather conditions, the crop is harvested with a combine when the plants are dry and the majority of seeds are mature. This practice leads to considerable loss from shattering and birds, and in other areas the crop is cut and placed in windrows to dry as soon as the plants start to dry out and the earliest seeds are mature. An approximate guide to this stage in a crop is when the later ripening plants start to become yellow. Stacks left to dry must be on sheets to avoid loss from shattering.

Threshing

Spinach seed from cut and dried plants is threshed with a small–drum thresher or a cereal thresher. In the latter case the recommended drum speed is about 700 rpm, and the concaves are set relatively wide to minimize the amount of broken stalks. In addition to separating the seeds from the plant, the threshing operation breaks up the clusters of seeds.

Seed yield

The generally accepted seed yield is approximately 800 kg/ha, although yields of up to 2000 kg are reported. The yield of hybrid cultivars per unit area is very similar to open pollinated cultivars even allowing for discarding the male lines.

1000 grain weight

The 1000 grain weight of spinach is approximately 10 g.

The main seed-borne spinach pathogens with common names of the diseases they cause

Pathogens	Common names
Cladosporium variabile (Cooke) de Vries, syn. *Heterosporium variabile*	Leaf spot
Colletotrichum dematium (Pers. ex Fr.) Grove f. *spinaciae* (Ell. and Halst.) Arx, syn. *C. spinaciae* Ell. and Halst	Anthracnose
Colletotrichum spinaciicola Chupp and Sherf	
Phyllosticta spinaciae Zimm.	Leaf spot
Rhizoctonia solani Kühn	Damping off
Verticillium dahliae Kleb.	
Verticillium sp.	Wilt

REFERENCES

Anon. (1981) *Guide to the Arrangements in North Essex for the Prevention of Injurious Cross-Pollination of Seed Crops of Allium (Onion), Beta (beet) and Brassicas.* Ministry of Agriculture. Fisheries and Food, London.

Chroboczek, E. (1934) A study of some ecological factors influencing seed-stalk development in beets (*Beta vulgaris* L.), *Cornell Univ. Agric. Exp. Sta. Mem.* **154**, 1–84.

Dark, S. O. S. (1971) Experiments on the cross-pollination of sugar beet in the field, *J. Nat. Inst. Agric. Bot.*, **12**, 242–66.

Sneep, J. (1958) The breeding of hybrid varieties and the production of hybrid seed in spinach, *Euphytica* **7**, 119–122.

FURTHER READING

Parlevliet, J. E. (1968) Breeding for earliness in spinach (*Spinacea oleracea*) as based on environmental and genetic factors, *Euphytica* **17**, 21–7.

8 COMPOSITAE

Although this family contains a large number of genera there are relatively few which are important as cultivated vegetables. The main vegetables which are produced from seed are:

Lactuca sativa L.	lettuce
Cichorium endivia L.	endive
C. intybus L.	chicory
Tragopogon porrifolius L.	salsify or oyster plant
Taraxicum officinale Wigg.	dandelion

In addition to the five species above there are *Helianthus tuberosus* L. (Jerusalem artichoke) and *Cynara scolymus* L. which are generally vegetatively propagated from selected clones.

All of the above vegetable crops are of more importance in the temperate regions than in the tropics. Lettuce, which is extremely important in the temperate regions especially Europe and North America, has gained in popularity in the tropical and arid regions of the world during recent years. Three other very important cultivated members of Compositae, although not vegetables, are *Carthamus tinctorius* L. (safflower) *Helianthus annuus* L. (sunflower) which are both important oilseed crops, and *Chrysanthemum cinerariaefolium* (Trev.) Bocc. which is grown for the production of the insecticide pyrethrum.

Locally important herbs, flavourings and drug plants include *Anthemis nobilis* L. and *Matricaria chamomilla* L. which are used for chamomile production, *Artemisia absinthium* L. (wormwood) and *A. dracunculus* L. (tarragon).

LETTUCE: *Lactuca sativa* L.

Origin and types

The cultivated lettuce *Lactuca sativa* is generally thought to be derived from the wild species of *L. serriola* L. There is a very wide range or variation within the cultivated forms of *L. sativa* which are divided into four types. The divisions, based on morphological characters are:

1. *L. sativa* var. *capitata* L., the cabbage or head lettuce which is generally subdivided into crispheads (iceberg) and butterheads. The crispheads have firm hearts produced by the close overlapping of coarse-veined brittle

leaves with prominent midribs. Typical cultivars are Great Lakes, New York (syn. Webb's Wonderful) and Chou de Naples.

The butterheads have relatively soft textured leaves with a greasy appearance. As in the crispheads the leaves of mature plants form a heart, although less firm than in the former type. Typical cultivars are White Boston, Cobham Green and Continuity.

2. *L. sativa* var. *longifolia* Lam., the cos or Romaine lettuce. The upright, relatively narrow crisp textured leaves form a closed head. Examples are Lobjoit's Green and White Paris.

3. *L. sativa* var. *crispa* L., the 'leaf' or curled lettuce does not form a heart but has a loose head of leaves. Some of the cultivars have very curly and fringed leaves, typical of which are Salad Bowl and Grand Rapids.

4. *L. sativa* var. *asparagina* Bailey (syn. var. *angustana* Irish); all the forms of this group have typical fleshy stems which are generally the main culinary attraction especially in Asia. A common cultivar of this type grown in Northern Europe and North America is Celtuce; some within the type have a light-grey leaf colour. The main morphological types of cultivated lettuce are shown in Fig. 8.1. Systems for classification and identification of lettuce cultivars have been documented by Bowring (1969) and Rodenburg and Huyskes (1964). Figure 8.2 shows the young leaf morphology of the different types. Classifications according to seedling characters only have been based on work by Rodenburg (1958) and while extremely useful for 'distinctness, uniformity and stability' tests, the characters are useful for selection or roguing only if observations are made before about the fifth leaf stage. Some of the seedling characters such as absence or presence of anthocyanin cannot be observed after the first true leaf has emerged.

Cultivar description of lettuce

Colour of seed   black, brown or white.

Type of head   cos, butterhead or crisp; hearting or leafy type

Season   summer outdoor, winter or protected crop; specific temperature regime if grown as a protected crop

Seedling and/or young plant characters at 3–4 leaf stage
  Anthocyanin pigmentation
    Leaf   colour: yellow, green or blue-green; degree of colour: light, medium or dark
                    shape: round, obvate
                    form: undulation and cupping
                    edge: degree of dentation

Mature plant characters
  Leaf   lustre
                    degree of blistering
  Time to maturity relative time plants will remain in satisfactory condition for market before bolting

Fig. 8.1 Main types of cultivated lettuce showing, (top) left to right, butterhead, crisp and cos (bottom).

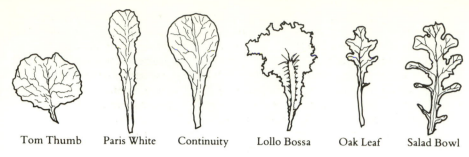

Tom Thumb   Paris White   Continuity   Lollo Bossa   Oak Leaf   Salad Bowl

Fig. 8.2   Third leaf of some lettuce cultivars showing range of morphological types.

Resistance to specific pathogens and their races, e.g. *Bremia lactucae* Regel (downy mildew) and pests, e.g. *Pemphigus bursarius* L. (lettuce root aphid).

Nutrition

This crop should be grown in a soil which has a pH of at least 6.0. In practice it is preferable to adjust the pH to 6.5 by adding a suitable liming material during preparation. This is because the lettuce crop is susceptible to calcium deficiency.

The lettuce seed crop responds to the same pattern of nutrients as the market crop and seed producers generally use a base dressing of 3 : 2 : 2 N : P : K. Nitrogen plays an important role in total seed yield which responds to up to 80 kg nitrogen per hectare. However, excess nitrogen tends to produce a loose atypical head which is difficult to confirm as true to type when roguing. Nitrogen can be applied during seed bed preparation but supplementary top dressings or foliar sprays (using urea) should be applied during the crop's growth. This is especially useful if there is likely to have been much leaching. Extra applications of nitrogen when the plants are bolting will increase seed yield.

Sowing

The seed is drilled into rows 50–60 cm apart at a rate of 1.5 to 2 kg per hectare. The young plants are thinned to between 25 and 30 cm apart within the rows. Transplants are planted 25 cm apart with 50–60 cm between the rows.

Irrigation

The lettuce seed crop responds directly to irrigation. In addition to increasing total plant weight in the vegetative stage a satisfactory supply of water assists in the increase of total seed yield, although there is a tendency for seed crops receiving frequent irrigation to be up to five days later in maturity. The significant gain in seed yield is generally considered to outweigh the possible disadvantage of a slight delay to optimum harvest time of the seed.

In areas where the total water requirement is applied through irrigation systems, growers generally maintain a drier regime after harvesting. Most lettuce seed producers reduce or cease irrigation from time of flowering. Overhead irrigation is detrimental once seed has begun ripening because the impact of the water droplets causes ripe seed to drop. In some areas irrigating late in the life of the crop encourages weed growth, and this can make harvesting difficult and increase contamination by weed seeds.

Flowering

The modern cultivars of lettuce are day-neutral (the summer types) or long-day plants (the types generally used for winter greenhouse production). The large amount of plant breeding work in the last two decades has produced some cultivars which are intermediate. It is therefore important to grow individual cultivars under the conditions prescribed by their breeder. Some cultivars which require long days will not heart under short days while cultivars which require short days for hearting will bolt before hearting if grown under long days.

Pollination

The inflorescence of lettuce, which is called a capitulum (see Fig. 8.3), contains approximately twenty-four individual florets. These are highly developed in favour of self-pollination and the crop is therefore largely self-fertilized. The mechanism of their self-pollination is shown in Fig. 8.4. However, some cross-pollination can take place between lettuce and also between cultivated lettuce and some wild *Lactuca* spp. For example, the cultivated lettuce *L. sativa* is cross-compatible with *L. serriola*, which is a wild plant and fence-line weed especially in countries around the Mediterranean and parts of Asia. If mechanical contamination of the lettuce seed crop with seeds of wild *Lactuca* occurs it presents insurmountable problems during the subsequent processing.

Isolation

For the production of certified seed a period of three years should have elapsed from a previous lettuce seed crop produced on the same site or two years from a previous market crop.

Although up to 5 per cent cross-pollination has been observed in lettuce in some areas most authorities regard it as a self-pollinating crop and only specify a physical barrier (e.g. adjacent sections of greenhouses) or a minimum of 2 m between different cultivars.

Fig. 8.3 Lettuce seed head showing range of inflorescence development from anthesis to mature seeds.

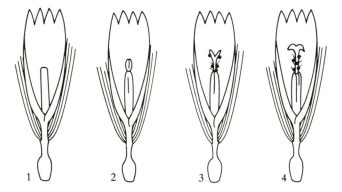

Fig. 8.4 Diagrammatic representation of the sequence (1,2,3,4) and self-pollination method in a lettuce floret (after Thompson 1938).

Roguing and selection

There are three main stages for roguing and selection, which are:

1. The young plant during the four- to six-leaf stage.
2. The mature plant at time of heading.
3. After bolting has commenced.

There is an additional earlier stage in the life of the plant when it is still a seedling. In practice this is only used when the plants are propagated prior to planting out, or if relatively few seedlings are being examined.

The most important stage is at the time of heading; in practice this is usually the only stage at which commercial seed crops are rogued.

The features used in the assessment of trueness to type of lettuce seed crops are mainly based on morphological characters observed during the vegetative stages of plant growth up to and including hearting.

Young plant characters

The general plant habit and other significant foliage characters can be assessed between the four- and six-leaf stage. The characteristic colour of a lettuce cultivar is seen best at this stage and this is a specially important time for the confirmation of leaf colour in the cultivars of the butterhead group. The crispheads are mainly dark green at all stages. In addition to leaf colour the following foliage characters can be examined:

(a) Anthocyanin: absent or present; if present seen in angular patches or generally distributed.
(b) Leaf margin: especially important for confirmation of leaf margin characteristics of butterheads during young plant stage. Both degree of undulation and pattern of indentation, if any, can be assessed as separate characters.
(c) Leaf shape: may range from spatulate to round; a distinctive character of any variety.
(d) Leaf cupping: individual leaves may be flat, cupped or inversely cupped.
(e) Leaf attitude: outer leaves are spreading, semi-erect or erect.
(f) Leaf texture: this can range from smooth to blistered. Intermediate degrees of blistering are difficult to assess although comparisons between plants of different cultivars can help in determining this character.

Mature plant characters

At this stage the general uniformity of a crop can be examined, and particular attention given to the time that individual plants reach potential market maturity, before being allowed to bolt for seed production.

The non-hearting rogues can be detected in hearting cultivars and firmness and size of heart can be assessed. Another important character is the length of time that a head will stand in the field once it has reached the marketable stage before it commences to bolt. The time taken to bolting is a cultivar characteristic, although it can vary with the environment and be influenced by the time of the year that the cultivar is grown.

The majority of leaf characters are also assessed at this stage. It is an ideal time for the determination of leaf colour and leaf margin characteristics in the crispheads. Additional characters observed at market maturity in all types are leaf overlap around the head, presence of anthocyanin, head shape, firmness and relative size.

Bolting plant characters

The form of the bolting plant can provide important evidence of trueness to type. Generally the habit of the terminal stem is taken into consideration at the same time as any branching, basal side shoots and height at flowering. The stem will be either fasciated or circular in transverse section, according to cultivar. The shape, colour and attitude of leaf bracts (or leaflets) below the inflorescences are useful characters. In addition, the state of leaf bracts assists in determining the plant's health at this stage; they will, for example, show lettuce mosaic symptoms if the virus is present.

Disorders

These include susceptibility to the different types of tipburn (marginal, heart or 'glassiness') which are regarded as physiological diseases. (The susceptibility to a specific pathogen can be assessed during roguing. These include lettuce downy mildew (*Bremia lactucae* Regel) and grey mould (*Botrytis cinerea* Pers. ex Pers.). Assessing resistance to specific races of pathogens is specialized work and is normally done during breeding programmes by close liaison between plant breeders and plant pathologists. In some areas of the world where lettuce root aphid (*Pemphigus bursarius* L.) is an important pest it is desirable to check susceptibility to attack, but as with downy mildew assessments it is normally only important in breeding programmes rather than during cultivar maintenance and multiplication.

It is stressed that plants showing mosaic or symptoms of other viruses should be removed at each roguing stage.

Lettuce mosaic virus (LMV)

This virus causes a yellow mottling of the leaves seen as a clearing between the fine veins when a portion of leaf is held up to the light. Plants which become affected relatively early in their life are usually stunted and frequently fail to heart (Fig. 8.5).

When seed is saved from infected plants 0–15 per cent of the seeds carry the virus in the embryo. Plants which are infected as a result of seed transmission act as reservoirs from which sap-sucking insect vectors transmit the virus to other plants in the same or neighbouring crops. The aphid *Myzus persicae* is especially important as a vector of this virus.

In addition to ensuring satisfactory isolation from other lettuce crops it is important to rogue out infected plants as soon as the symptoms are observed. Basic seed stocks can be produced in insect-free structures or, in some parts of the world such as Australia and California, at temperatures too high for aphid attack. There are other viruses that cause mosaic-like symptoms but they are not seed-borne.

Mosaic-indexed seed

This is sometimes referred to as mosaic-tested seed. It is a lettuce seed stock which has been shown to have less than 0.1 per cent of seeds infected with LMV (i.e. when grown on less than one plant in a thousand seedlings will be infected with LMV from seed transmission).

Fig. 8.5   The plant on the left was grown from a seed infected with lettuce mosaic virus, the plant on the right is normal.

Mosaic-indexed seed is produced under strictly controlled conditions which incorporate satisfactory isolation, roguing and virus vector control. The resulting seed is then subject to mosaic indexing.

When seed is mosaic-indexed a sample of at least 5000 seedlings is grown on from the harvested seed crop. The seedlings are usually raised and grown in a cool insect-proof greenhouse or structure. Any young plants which have been infected from the seed will show the typical mosaic symptoms by the fourth or fifth leaf stage.

Seed samples which have been found to produce less than 0.1 per cent seedlings with LMV are accepted as 'mosaic-indexed'. The production and use of indexed seed greatly assists the market producer and assists the seed grower to obtain higher yields of lettuce seed.

Seed stalk emergence

The seed stalk does not always emerge readily from the lettuce heart. Although this is not a major problem with the butterheads it can be with the crispheads because they have very tight and firm hearts which are mechanical barriers. The poor emergence results in the development and elongation of basal shoots, relatively few flowering shoots on the main axis, which becomes distorted, and a reduction of the potential seed yield. The general tardiness of bolting of firm-hearted lettuce also results in an increase in the mature plant's exposure to pathogens such as *Botrytis cinerea*.

Traditionally, several mechanical methods have been used to assist emergence and elongation of flower shoots. These have included deheading, slashing and quartering of the heads with special tools to assist the breakage of the solid heart without damaging the unexposed growing point; in England this operation is called 'racing'. Another simple method is to give a sharp blow with the palm of the hand to the top of the lettuce head. This fractures the leaf bases leaving the growing point unharmed. An alternative is manually to peel off the leaves around the heart, but this is only really possible if relatively few plants are being seeded. The timing of the mechanical or manual operations is important, otherwise the extended flowering shoot inside the heart will be damaged.

Growth-regulating chemicals have been used to promote bolting in lettuce. Gibberellic acid has been applied to lettuce before hearting to promote early bolting, thus avoiding the problems associated with shoot emergence. The application of aqueous solutions of gibberellic acid at concentrations between 20 and 500 ppm has been recommended for butterheads. The application of gibberellic acid solutions before hearting does not permit roguing at hearting time.

Applications of gibberellic acid at hearting time, while assisting flower head emergence of the butterheads, did not improve flower head emergence from hearted crispheads (George and Jamieson 1980).

Seed harvesting

The succession of inflorescences during anthesis subsequently provides a steady sequence of ripe seed. The length of time from flowering to ripe seed produced on an individual capitulum is between twelve and twenty-one days, depending on the environment. High temperatures speed up this relatively fast rate of seed development.

If the seed producer waits for the development and ripening of the seeds from the later inflorescences, the earlier ripened ones will probably have been lost. It is therefore general practice to cut and harvest the seed when an estimated 50 per cent of seed heads are ready on a typical sample plant. The stage of ripeness at which the pappus is fully developed and dry is referred to as 'feathering'. In Fig. 8.6 all the former inflorescences on the plant have 'feathered'.

The standing lettuce seed crop can be hand or machine cut and put in windrows. A machine which causes the minimum of shattering should be used for the cutting process. There is less shattering if the plants are cut with dew on them. The cut material is normally left in windrows for up to five days, but in very arid areas the seed is extracted the same day as cutting. After windrowing the seed is extracted in stationary threshers or passed through a combine. If the seed is combined direct from the standing crop many small pieces of wet plant debris, such as fragments of leaf bract, will be mixed with it; this causes an undesirable increase in seed moisture unless the material is dried quickly.

In production areas with plenty of available labour, the seed can be harvested from single plants by shaking their heads into a canvas bag or sack. If this is done every two or three days the maximum seed yield is collected.

Fig. 8.6   Lettuce plants in which all the inflorescences have 'feathered'.

Seed cleaning

Pre-cleaning vegetable seeds of Compositae is an important operation because the weight of chaff in mechanically harvested seed can be as high as one and a half times the weight of the seed itself. The initial cleaning can be done with an air-screen machine. There is usually some plant material such as small pieces of flower stem remaining in the sample; these can be removed by passing the seed through a disc separator or an indent cylinder.

Seed yield

A satisfactory seed yield for hearting lettuce under good conditions is between 0.5 and 1 tonne per hectare.

Seed weight

The 1000 grain weight of lettuce seed is from 0.6 to 1.0 g according to cultivar.

The main seed-borne lettuce pathogens with common names of the diseases they cause

Pathogens	Common names
Marssonina panattoniana (Berl.) Magnus	Ring spot, anthracnose
Septoria lactucae Pass syn. *Aschochyta lactucae* Rostrup	Leaf spot
Pseudomonas cichorii (Swingle) Stapp	Leaf blight
Viruses	Arabis mosaic virus
	Lettuce mosaic virus
	Tobacco ring spot virus

WITLOOF CHICORY: *Cichorium intybus* L.

This is a crop grown as a winter vegetable by specialist producers especially in Northern Europe. The crop is produced by forcing the one-year-old dormant roots.

The cultivars are normally classified according to their relative earliness when forced as there are very few other clear distinctions. The time to maturity of a specific cultivar depends on its vernalization requirement. Rodenburg and Huyskes (1964) have developed a scheme for classifying Witloof chicory according to the length of the core, i.e. the stem in the head. They found that the longer the core the earlier the cultivar. This is a correlation with the vernalization requirement, as the successive cultivars have an increasing cold requirement. The cultivars with the lower cold requirements will tend to bolt when sown relatively early in their first year.

Cichorium intybus is also used for the production of roots which are dried, ground and used as an additive to coffee. There are no records of cultivars specially developed for this use.

Agronomy and seed production

The seed crop is grown as a biennial and subsequently treated similarly to lettuce. Some Dutch and French seed companies have developed F_1 hybrid cultivars of Witloof chicory.

This crop is largely cross-pollinated and there should be a minimum of 1000 m between seed crops of different cultivars.

Seed yield

The average chicory seed yield is approximately 1 tonne per hectare.

Seed weight

The 1000 grain weight of chicory seed is *c.* 1.4 g.

The main seed-borne chicory pathogens with common names of the diseases they cause

Pathogens	Common names
Alternaria cichorii Nattr., syn. *A. dauci* (Kühn) Groves and Skolko f. sp. *endiviae* (Nattr.) Janezic	Black leaf spot
Gibberella avenacea Cook, syn *Fusarium avenaceum* (Fr.) Sacc.	
Rhizoctonia solani Kühn	
Virus	Chicory yellow mottle virus

ENDIVE: *Cichorium endivia* L.

This is a popular salad crop in Northern Europe, especially France and Italy where seed companies offer a range of cultivars. Some of the traditional market gardens grow the plants to maturity and then 'blanch' them by covering with inverted pots or boxes. The present-day cultivars have been developed for production and marketing in a similar way to lettuce, i.e. without 'blanching'.

Origins and types

It is not clear if the cultivated derivatives of this species are of Indian or Mediterranean origin but selections of the species have been cultivated in both areas for centuries (Hedrick 1972). Some forms of *C. endivia* are annuals but seeds of the main cultivars offered by continental seed companies are produced as biennials.

Cultivar description of endive

Season   relative time of maturity, summer or autumn type; suitability for protected cropping

Size   relative size of mature head

Leaf   colour: green, degree of white centre to mature head
shape: broad leaved type or curled (Fig. 8.7)

Fig. 8.7 Broad leaved endive type (left) and curled leaved type (right).

Agronomy

The seed crop has the same soil requirements as lettuce. Seeds are sown in the late summer or autumn in frost-free areas. As the plants are more vigorous than lettuce when flowering, lower plant densities are used per unit area, with rows 75 cm apart and plants thinned to between 25 and 50 cm apart according to the vigour of the cultivar.

Roguing stages

The plants are examined as described for lettuce during the rosette and mature heart stages.

This species is considered to be mainly self-pollinating and the isolation requirements are therefore similar to lettuce, but different types should be isolated by a minimum of 50 m.

The two main roguing stages are:

1. Young plants in rosette stage. Remove plants which are bolting early. Check the general morphology and leaf characters.
2. Market maturity stage. Remove plants bolting early. Examine for tolerance to adverse weather, especially autumn-maturing cultivars.

Seed yield

The average endive seed yield is approximately 500 kg per hectare, although good yields in Italy reach 1 tonne per hectare.

Seed weight

The 1000 grain weight of endive seed is *c.* 1.3 g.

Seed-borne pathogens

The main seed-borne pathogen of endive is *Alternaria cichorii* Nattr. syn. *A. dauci* (Kühn) Groves and Skolko f. sp. *endiviae* (Nattr.) Janezic (black leaf spot).

REFERENCES

Bowring, J. D. C. (1969) The identification of varieties of lettuce (*Lactuca sativa* L.), *J. Nat. Inst. Agric. Bot.* **11**, 499–520.

George, R. A. T. and Jamieson, C. E. (1980) Investigations into the use of gibberellic acid for seed production of hearted lettuce, *Proceedings Eucarpia Meeting on Leafy Vegetables*, pp. 26–29. Glasshouse Crops Research Institute, England.

Hedrick, U. P. (ed.) (1972) *Sturtevant's Edible Plants of the World.* Dover publications, New York.

Rodenburg, C. M. (1958) The identification of lettuce varieties from the young plant, *Euphytica* **7**, 241–6.

Rodenburg, C. M. and Huyskes, J. A. (1964) The identification of varieties of lettuce, spinach and Witloof chicory, *Proc. Int. Seed Test. Ass.* **29(4)**, 963–980.

FURTHER READING

Bukovac, M. J. and Wittwer, S. H. (1958) Reproductive responses of lettuce (*Lactuca sativa* L. variety 'Great Lakes') to gibberellin as influenced by seed vernalization, photoperiod and temperature, *Proc. Amer. Soc. Hort. Sci.* **71**, 407–11.

Dunn, J. A. (1960) Varietal resistance of lettuce to attack by the lettuce root aphid, *Pemphigus bursarius* L., *Ann. Appl. Biol.* **48(4)**, 764–70.

Eenink, A. H. (1980) Breeding research on Witloof chicory for the production of inbred lines and hybrids, *Proceedings Eucarpia Meeting on Leafy Vegetables*, pp. 5–11. Glasshouse Crops Research Institute, England.

Globerson, D. and Ventura, J. (1973) Influence of gibberellins on promoting flowering and seed yield in bolting-resistant lettuce cultivars, *Israel J. Agric. Res.* **23**(2), 75–7.

Grogan, R. G. (1981) Control of lettuce mosaic with virus-free seed, *Plant Disease* **64**(5).

Harrington, J. F. (1960) The use of gibberellic acid to induce bolting and increase seed yield of tight-heading lettuce, *Proc. Amer. Soc. Hort. Sci.* **75**, 476–9.

Hawthorn, L. R. and Pollard, L. H. (1958) Production of lettuce seeds as affected by soil moisture and fertility conditions, *Utah Agric. Exp. Sta. Bull.* **386**, 4–23.

Hiraoka, T. (1967) Ecological studies on the salad crops I. Effects of temperature, photoperiod and gibberellin spray on bolting and flowering time in head lettuce, *J. Jap. Soc. Hort. Sci.* **36**, 70–8.

Huyskes. J. A. (1962) Cold requirements of Witloof chicory varieties (*Cichorium intybus* L.) as a yield-determining factor, *Euphytica* **11**, 36–41.

Lindqvist, K. (1960) On the origin of cultivated lettuce, *Hereditas* **46**, 319–50.

Lovato, A. and Montanari, M. (1981) The viability of carrot and chicory seed as affected by desiccant sprays, *Acta Horticulturae* **111**, 175–81.

Maxon Smith, J. W. (1976) Nutritional effects on glasshouse lettuce seed parent plants and their progeny, *Hort. Res.* **16** 45–52.

Montanari, M. and Lovato, A. (1981) The yield and quality of carrot and chicory as affected by desiccant sprays, *Acta Horticulturae* **111**, 167–73.

Olivieri, A. M. and Parrini, P. (1980) Some breeding aspects in Italian types of *Cichorium* spp., *Proceedings Eucarpia Meeting on Leafy Vegetables*, pp. 12–25. Glasshouse Crops Research Institute, England.

Soffer, H. and Smith, O. E. (1974) Studies on lettuce seed quality. V. Nutritional effects, *J. Amer. Soc. Hort. Sci.* **99**, 459–63.

Thompson, R. C., Whitaker, T. W., Bohn, G. W. and Van Horn, C. W. (1958) Natural cross-pollination in lettuce, *Proc. Amer. Soc. Hort. Sci.* **72**, 403–9.

9 CRUCIFERAE

This family contains several important vegetables and includes:

Brassica oleracea L.	the brassicas or cole crops
var. *acephala* DC	kale
var. *capitata* L.	cabbage
var. *botrytis* L.	cauliflower
var. *italica* Plenck	sprouting broccoli
var. *gemmifera* Zenker	Brussels sprouts
var. *gongylodes* L.	kohlrabi
Brassica campestris L.	turnip and related crops
subsp. *rapifera*	true turnip
subsp. *chinensis* Jusl.	pak-choi or Chinese mustard
subsp. *pekinensis* (Lour.) Rupr.	pe-tsai or Chinese cabbage
Brassica napus L.	swede
Brassica juncea (L.) Czern. and Coss.	brown mustard, Indian mustard
Brassica nigra (L.) Koch	black mustard
Sinapis alba L. (Syn. *B. alba* (L.)) Rabenh.	white mustard
Lepidium sativum L.	garden cress
Raphanus sativus L.	radish
Rorippa nasturtium aquaticum (L.) Hayek	watercress

COLE CROPS: *Brassica oleracea* L. vars.

Examples of the different types of *Brassica oleracea* are given in Figs 9.1 and 9.2.

The general history of these species was discussed by Thompson (1976) and detailed information on each of the cultivated types was provided by Nieuwhof (1969).

The seed production methods for cabbage, cauliflower, Brussels sprouts and kohlrabi are described below; sprouting broccoli, calabrese and the vegetable kales are not discussed here because they are relatively similar to these four main cole crops.

(a)

(b)

(c)

(d)

Fig. 9.1   Different types of *Brassica oleracea*: (a) Cauliflower, (b) Savoy cabbage, (c) Brussels sprouts and (d) Kale.

Cultivar descriptions of the main cole crops

Cabbage: *Brassica oleracea* L. var. *capitata* L.

Method of seed production	open pollinated or F_1 hybrid, type of hybrid
Season and use	fresh market, processing or storage
Morphological type	head shape: round, flat or pointed
Outer leaves	colour: green, blue-green, red or 'white' pigmentation degree of waxiness of cuticle texture (e.g. blistered)

Resistance to specific pathogens, resistance to splitting and early bolting:

(Some cultivars are susceptible to rosetting, i.e. forming open heads rather than tightly enclosed heads and in addition to the above information some cultivar descriptions report the normal incidence of this.)

Cauliflower: *Brassica oleracea* L. var. *botrytis* L.

Season of production, suitability for overwintering, frost resistance, cold requirement (if any) for curd initiation

Leaf	shape
	characters of midrib and veins
	protection to curd provided by leaves
Curd	colour: green, white, pink
	relative size
	pattern of florets in curd, if characteristic of the cultivar
Flower	colour

Resistance to specific pathogens, e.g. *Mycosphaerella brassicicola* (ring spot).

Brussels sprouts: *Brassica oleracea* L. var. *gemmifera* Zenker

Open pollinated or F_1 hybrid, type of hybrid

Season of production, suitability for specific market outlet (e.g. freezing or fresh market)

Suitability for mechanical harvesting

General vegetative characters	height
Leaf	shape, colour
	petiole length
Sprouts ('buttons'):	relative size
	colour

Resistance to specific pathogens, e.g. *Mycosphaerella brassicicola* (ring spot)

Kohlrabi: *Brassica oleracea* L. var. *gongylodes* L.

Season and uses, suitability for protected cropping, resistance to early bolting

'Bulb' (i.e. swollen stem) characters:

colour	white, red, purple
	extent of pigmentation if speckled
shape	globe or relatively flat
internal quality	degree of fibre
Foliage	short or tall, pose of leaves
	relative length of petioles

Typical morphological differences of kohlrabi are shown in Fig. 9.2

Fig. 9.2 Typical morphological types of kohlrabi.

Nutrition and pH

The optimun pH for cole crops is 6.0–6.5 and any required lime should be applied during soil preparation. Acid soil conditions not only affect the availability of micro-nutrients such as molybdenum, but also increase the incidence of club root (*Plasmodiophora brassicae* Woron.).

The N : P : K ratio applied during preparation varies widely between different seed production areas, but the general recommendation is 1 : 2 : 2. High nitrogen levels result in 'soft' plants less able to withstand frost, and for this reason top dressings of nitrogen are applied in the spring. The spring application also replaces nitrogen leached during the winter. Where deficiencies of available manganese, boron and molybdenum exist, suitable additions of the appropriate materials should be applied. Cauliflowers are especially susceptible to molybdenum deficiency which causes the condition known as 'whiptail'. A solution of sodium molybdate is applied to the young plants prior to planting out in areas where the deficiency is likely to occur. Boron deficiency causes 'hollow-stem' or corky patches on the stems of cole crops. The deficiency in cauliflowers can also be responsible for the browning of parts of the curds (Fig. 9.3).

Irrigation

Most references in the literature relating to the irrigation of cole crops refer to their vegetative stage. Increasing the available water when the plants are in the vegetative stage will increase the amount of foliage produced but no work has been reported on the effects of water availability on seed yield.

Fig. 9.3   Boron deficiency symptoms showing in a cauliflower curd.

However, work at the National Vegetable Research Station (Winter 1974) showed that cauliflower curd size is significantly increased by supplementary irrigation three weeks before the crop is normally harvested. Water stress in the cole crops increases the thickness of the waxy cuticle and produces plants with a relatively blue-green coloration. This frequently occurs under tropical or arid conditions and can increase the difficulty of confirming some foliage characters during roguing.

Plant production, sowing and spacing

Cabbage, cauliflower and Brussels sprouts plants are normally raised in seed beds and planted out. The sowing dates depend on the local custom and experience with specific crops. Sowing is timed so that the cultivar receives sufficient vernalization (if required), and also to allow for the plants to survive hard weather and flower and seed in suitable weather conditions. Sixty g of seed thinly sown in drills 20 cm apart to produce approximately 10 seedlings per 30 cm will produce *c.* 10 000 plants, subject to satisfactory seedling emergence. Some of the specialized cauliflower seed producers sow cauliflowers in a greenhouse and prick off the seedlings into 7 cm pots or soil blocks.

Young plants from the seed beds, pot or soil blocks are planted out when they have reached the five to seven leaf stage. Obvious off-types, blind plants and any showing signs of serious pathogens are discarded at this stage.

The final spacing depends on the vigour of the cultivar, but a useful criterion is to ensure that the plant density is as high as possible but there should be sufficient space to allow inspectors to examine the crop. Rows are usually 60 cm apart, although some of the smaller cabbages are planted in rows 40 cm apart. The distance between plants within the rows is usually 6 cm. In some seed production areas a 'bed' or 'set' system of up to four rows is used with a wider space between the beds.

Some seed producers earth-up cabbage and Brussels sprouts plants after they are established in their final stations. This is done as part of the cultivation programme to control weeds but it also stops many plants toppling over when they are heavier.

Kohlrabi is usually direct drilled for seed production as a 'seed to seed' system, although transplanting is used for stock seed production. Unlike the other cole crops discussed above it is possible to vernalize the seed of some quick bolting cultivars of kohlrabi. The treatment is given before a spring sowing: the seeds are pre-soaked with water for approximately eight hours, then spread on moist sheets of filter paper at 20–22 °C. Between 70 and 90 per cent germination occurs in twenty-four hours at these temperatures. The moist seeds are stored at -1 °C for 35–50 days (Nieuwhof 1969). After thawing slowly the seeds are sown direct into the field. The use of this technique does not allow for roguing against early bolters.

Cultural techniques

Cabbage

If seed is to be produced from hearted cabbage it is usually necessary to incise the mature heads after checking trueness to type. There are several ways of achieving this but all have the objective of allowing the flower stalk to emerge unhindered by the mechanical barrier of the tightly folded leaves which encase it. A quick method is to cut the top of the head in the form of a cross, but the growing point must not be damaged. In Northern Europe the stems of winter cabbages are earthed-up to protect them from severe frost.

Cauliflower

When the curds are formed at a time of year with the expectation of subsequent fine weather for seed production to take place, as with early summer cauliflower, the plants are left *in situ*. Techniques for dealing with selected plants at other times of the year are discussed under basic seed production techniques.

Brussels sprouts

The terminal growing points are removed from plants after the final roguing. This encourages the development of flowering shoots from the lateral buds (i.e. 'sprouts) thereby increasing total seed yield and uniformity of seed maturity. Some seed producers remove some of the lower sprouts.

Kohlrabi

Overwintered crops are either left in the ground and protected from frost by earthing over or covering with a suitable material, or they are lifted and stored in frost-free conditions. The stored roots are re-planted in the early spring.

Flowering

Flower initiation in most cole crops is dependent on a low temperature stimulus. This requirement varies between the different types and between cultivars of individual types, which are usually regarded as biennials. The only exceptions are the summer cauliflowers which do not have a vernalization requirement and are considered to be annuals. The plant's age and leaf number are also important in its receptiveness to low temperature. The reader is referred to 'Further Reading' for research papers on this topic.

There is a great deal of interest in warmer areas of the world in producing seeds of some of the cole crops at higher altitudes by the application of gibberellic acid (GA_3). Work in Kenya by Kahangi and Waithaka (1981) showed that the incidence of flowering of some cabbage and kale cultivars at an altitude of 1941 m was increased if GA_3 was applied from one to two months after planting out.

Pollination

Most cole crops are predominantly cross-pollinated although some self-pollination can occur. The level of self-pollination is greatest in the summer cauliflower (Watts 1980). Bees and Diptera pollinate *Brassica* spp. Work by Faulkner (1962) showed blowflies to be very efficient when plants are flowering in confined places such as greenhouses.

All types of *Brassica oleracea* can cross with each other and it is extremely important to appreciate this when planning and checking their isolation for seed production. The types of *B. oleracea* grown as fodder and forage crops, such as cattle cabbages, marrow stem kale and thousand head kale, are also cross-compatible with the types grown as vegetables.

Isolation

Most authorities consider it important to have a greater recommended distance (up to 1500 m) between different types of *B. oleracea*, e.g. cabbages and cauliflowers, than between different cultivars of the same type, e.g. two cabbage cultivars (up to 1000 m). Some countries have introduced zoning schemes for some of the *Brassica* spp.

Roguing stages

Cabbage roguing stages

1. *Planting out, or before heading*   Check for general foliage characters.

2. *When heads have formed on most plants in the crop*:   Check head characters, including shape, relative size, firmness. Discard plants which are either too early or too late for the type.

Cauliflower roguing stages

1. *Before normal curding period*   Reject precocious or button curds which develop before the normal maturity period of the cultivar. Foliage – check that pose of leaves is true to type. Leaf number – shape and extent of crinkle according to type.

2. *Curd at market maturity*   Curd – colour, absence of bracts, absence of riciness, solidity (loose curds are undesirable). Shape of curd and leaf protection according to the cultivar.

Brussels sprouts roguing stages

1. *Planting out, or before significant buttoning commences*   Check for general plant vigour, relative height and foliage characters including colour.

2. *When the lower 'buttons' or buds reach normal market maturity*   Check general plant characters, productivity and quality of 'buttons'. Discard plants which are not within the maturity period of the cultivar.

Kohlrabi roguing stages

1. *During final thinning*   Remove any plants which are bolting early, also plants which are not true to type for colour, vigour or leaf characters.

2. *At normal market maturity of 'bulbs'*   Remove early bolting plants, check that 'bulb' shape, foliage and coloration are true to type.

Hybrid seed production

The ratio of pollinator to female rows is 1 : 2, although this will depend on the flowering ability of each parent and on whether or not the seed from each parent is to be saved.

One problem in F_1 hybrid seed production of cole crops, especially Brussels sprouts, has been the unacceptably high number of sibs produced. A sib is seed derived from self-pollination. The problems relating to F_1 hybrid seed production were discussed by Faulkner (1978). Other references relating to hybrid seed production of the cole crops are given in 'Further Reading'.

Harvesting

All of the cole crops have a strong tendency to shattering of their seed pods. It is therefore important that appropriate action is taken before seeds are lost.

As the seeds ripen the plants start to dry out and the orange-brown colour of the plant is the best sign of this. The approaching maturity of the majority of seeds on a plant can also be confirmed by opening a sample of the oldest pods which will be the first to become brown.

Many seed producers prefer to cut the ripening material by hand and place in windrows or on sheets to continue drying before extracting the seeds in stationary threshers. Direct combining can only be done in dry conditions and care must be taken to minimize the loss from shattering.

Threshing

The seeds of *Brassica* spp. crack very easily and it is therefore important that a relatively slow cylinder speed, not normally exceeding 700 rpm, is used although faster speeds may be required if the material is not very brittle.

Split seeds can be separated out from a seed-lot by a spiral separator.

Seed yield and 1000 grain weight

The reported seed yields vary from one area to another and also depend on the type but the following figures are a guide.

Cabbage	yield 700 kg/ha; 1000 grain weight 3.3 g
Cauliflower	yield 400 kg/ha; 1000 grain weight 2.8 g
Brussels sprouts	yield 600 kg/ha; 1000 grain weight 2.8 g
Kohlrabi	yield 700 kg/ha; 1000 grain weight 3.2 g

The main seed-borne pathogens of *Brassica* spp. with common names of the diseases they cause

Pathogens	Common names
Alternaria brassicae (Berk.) Sacc.	Grey leaf spot
Alternaria brassicicola (Schw.) Wilts., syns. *A. oleracea* Milbraith, *A. circinans* (Berk. and Curt.) Bolle	Black spot, wirestem
Ascochyta oleracea J. W. Ellis	Leaf spot
Leptosphaeria maculans (Desm.) Ces. and de Not., syns. *Phoma lingam* (Tode ex Fr.) Desm., *Plenodomus lingam* (Tode ex Fr.) Höhnel	Dry rot, black leg, black rot

Pathogens	Common names
Mycosphaerella brassicicola (Duby) Lindau, syns. *Asteromella brassicae* (Chev.) Boerema and van Kesteren, *Phyllostica brassicicola* Grove	Black ring spot
Plasmodiophora brassicae Woron	Club root
Pseudocercosporella capsellae (Ell. and Ev.) Deighton, syn. *Cercosporella brassicae* (Fautr. and Roum.) Höhnel	White leaf spot
Rhizoctonia solani Kühn	
Sclerotinia sclerotiorum (Lib.) de Bary	Watery soft rot, drop, white blight
Pseudomonas maculicola (McCulloch) Stevens	Bacterial leaf spot
Xanthomonas campestris (Pammel) Dowson	Black rot

Basic seed production

Production of spring cabbage seed

Three years are required from sowing the stock seed until harvesting the basic seed. A further two years are required before the morphological characters of the seed product can be verified in field trials and the basic seed used for multiplication to produce further generations, e.g. certified seed. The following scheme is used in the UK.

Year 1: Stock seed is sown in July and the young plants transplanted into the field in September.
Year 2: During May plants are selected for optimum cultivar characters and quality. The growing point and heart are removed from selected plants which are either left in the ground or transferred to greenhouses (polythene tunnels have proved very successful).
Year 3: The selected plants flower during late spring and early summer and pollinating insects are supplied if in greenhouses or other structures. Ripe seed is harvested during August. After threshing, cleaning and short-term storage the seed is either used as basic stock for further seed production or stored for a longer period while samples are grown on for field assessment, in which case samples are sown during year 4 and the sample, or progeny plots, assessed during year 5. (A year can be gained by allowing the selected plants to flower in the second year by not removing their growing points.)

Production of cauliflower basic seed

There are several techniques which are used to produce basic seed from selected cauliflower plants. In all cases the mother plants are grown in their normal season and final selections are made after confirmation of plant characters and curd quality. When environmental conditions in the field are expected to remain favourable for further flower development, anthesis and seed maturity the selected plants can be left *in situ*. This, for example, is the normal practice in Northern Europe and North America for the early summer cauliflowers. However, in the case of cultivars with curds maturing in the late summer, autumn and winter the selected plants are unlikely to survive if left *in situ*.

Careful lifting of selected plants and transferring them to a greenhouse is not always successful and techniques to produce seed from propagules of selected plants have therefore been developed. Watts and George (1963) reviewed the wide range of vegetative propagation methods used and also described a method of grafting pieces of selected curd on to specially produced cauliflower rootstocks. This method was further developed by Crisp and Lewthwaite (1974) who found that a suspension of 50 per cent a.i. benomyl applied in $500M$/water to each pot of grafted plant, with autumn cauliflower used as a root-stock, greatly improved the success rate and seed yield. Other methods of vegetative propagation include removing the stumps of selected plants to greenhouses and rooting the stem shoots which subsequently develop (Anon. 1980). References to vegetative propagation techniques are given in Further reading.

The adoption of vegetative propagation techniques for cauliflower seed production has led to the establishment of clones, but there has been a build-up of viruses in the clonal material as a result of successive vegetative propagation. Turnip mosaic virus and cauliflower mosaic virus were found to be present in many of the cauliflower clones tested by Walkey *et al.* (1974) who described a method for producing virus-free cauliflowers by tissue culture.

Selection against bracting
A cauliflower curd consists of many compressed and branched peduncles with thousands of pre-floral meristems (Crisp *et al.* 1975a). When the bracts grow through the curd surface the market value of the curd is reduced. Plants are selected against bracting during roguing or selection operations: However, because the incidence of bracting in any given cultivar is influenced by the environment, the same cultivars may display different degrees of the bracting defect in different years and production areas. Crisp *et al.* (1975a) described a two-tier system for selecting on macroscopic characters in the field, followed by approximately four weeks of aseptic culture during which different selections can be graded according to size of bracts.

Selection against undesirable curd colours
Although the generally accepted cauliflower curd colour is white or creamy white, individual plants occur in some cultivars which have a degree of pigmentation resulting from anthocyanin, carotenoids or chlorophyl. Coloured off-type curds in white curded cultivars are normally rogued out during the roguing operation when the curd quality is observed. Crisp *et al.* (1975b) described a technique for improving the selection against the purple colour defect. It involves aseptically culturing pieces of curd by the method described by Crisp and Walkey (1974) and visually inspecting the plant material for intensity of coloration after five weeks.

Production of basic Brussels sprout seed

Plants selected for basic seed production are either left *in situ* in the field or are transplanted into cages or greenhouses, in which case it is necessary to mass pollinate them. Blowflies have been found to be the most useful natural pollinating agent (Faulkner 1962). Work by Johnson (1958) has shown that progeny testing of Brussels sprouts can contribute very significantly to the open-pollinated cultivars.

RADISH: *Raphanus sativus* L.

Origin and types

Radishes have been cultivated for many centuries and inscriptions on the pyramids indicate that they were a food crop in *c.* 2000 BC (Becker 1962). Banga (1976) discussed the history and development of the four present-day cultivated radish types, which are:

1. *Raphanus sativus* var. *radicula*, cultivated for its swollen hypocotyl as a field and protected crop, mainly in temperate areas of the world, although they are cultivated in other areas too.
2. *Raphanus sativus* var. *niger*, the larger rooted type, mainly cultivated in Asia but still locally important in Germany.
3. *Raphanus sativus* var. *mougri*, which has a relatively insignificant root but is cultivated as a vegetable in South-East Asia for its edible foliage and relatively long seed pods.
4. *Raphanus sativus* var. *oleifera*, the type cultivated as a fodder crop, especially in Northern Europe where it has gained importance in the last decade.

All of these types described above readily cross-pollinate with each other and with the four wild *Raphanus* species, *R. raphanistrum*, *R. maritimus*, *R. landra* and *R. rostatus*. The occasional purple rooted off-types found in seed stocks are either a result of cross-pollination between cultivated and wild species or an admixture of seed from wild types which were not rogued out during seed production.

The main type from which seed is commercially produced on a large scale is *R. sativus* var. *radicula*. The cultivars of this variety are classified according to morphological characters including the colour and proportions of hypocotyl, and leaf size and number at market maturity. Physiological characters including earliness, time to bolting and duration of mature hypocotyl before onset of pithiness are also included. Fifty-five cultivars recommended for protected cropping were classified into twenty-five classes by George and Evans (1981).

Cultivar description of radish

Season of use, suitability for specific day-length, vernalization requirement for bolting (if any), suitability for early production as a protected crop.

Root (i.e. swollen hypocotyl)	shape, globe, oval intermediate or long relative size colour, red, white, yellow, black or bi-colour, if bi-colour the proportions of the two colours are defined
Foliage	number of leaves when mature relative height of leaves leaf shape
Flower colour	(This character is not normally stated except as additional information for seed producers.)

Agronomy

Soil pH and nutrients

Radish crops tolerate slightly acid conditions and a soil pH between 5.5 and 6.8 is suitable. The general N : P : K fertilizer application is a ratio of 1:3:4 although a lower amount of K is applied where there are satisfactory residues.

Seed production systems

Both 'root to seed' and 'seed to seed' systems are used for radish seed production. The 'root to seed' is used for the biennial types especially in Europe where the roots are lifted, the tops taken off and stored, usually in clamps, during the winter. It is also the method used for stock seed production of the annual types but in this case the material is re-planted immediately after selection.

The 'seed to seed' system is used for final multiplication stages where inspections of the mature root are not considered necessary and is normally used only for spring sown seed crops unless the cultivar has a vernalization requirement.

Sowing rates and spacing

Sowing rates of up to 6 kg per hectare are used, the seeds are drilled in rows from 50 to 90 cm apart to give a final plant distance within the rows of approximately 5 to 15 cm.

Flowering

According to Banga (1976) there has been selection for adaptation to different day-lengths and type of seasons. The cultivars developed for early spring production have an annual habit, and material without a vernalization or specific day-length requirement has been developed from both *R.s. radicula* and *R.s. niger*. However, these types flower earlier when grown in long-days.

Selection of types suitable for crop production in summer, autumn or winter has resulted in cultivars which are biennials with a vernalization requirement. The removal of early bolting plants from late spring and summer cultivars has probably resulted in the decreased sensitivity to daylength (Banga 1976).

The flowers are cross-pollinated by bees and some other insects.

Isolation

The recommended isolation distance is 1000 m, although in some countries the distance between commercial seed stocks of similar cultivars may be 200 m.

Roguing stages

1. (root to seed system) At market stage of roots.
   Root size: shape, colour, proportions of each colour on bi-coloured cultivars, solidity.
   Leaves: number, size, shape.

2. At stem elongation.
   Remove early bolting plants and off-types according to stem colour.
   Remove wild radish types.
   Check that remaining plants are true to type for foliage and stem characters

3. At flower bud
   As described in stage 2.

Basic seed production

The root to seed system is used for basic seed production. When selecting plants according to their external morphology, care should be taken to ensure that those with pithy roots are rejected. Traditionally, selected roots were examined for internal solidity by removing a small wedge of tissue with a knife, but Watts (1960) described an immersion technique which is very suitable for screening roots for solidity at or after normal market maturity and can be used to increase the length of time that they remain solid. After selection the plant's leaves are twisted off (leaving the growing point undamaged) and the roots put in a bucket of water. Those roots with a degree of pithiness float and are discarded. The solid roots sink and are retained. The selected solid roots are then planted up and grown-on for seed production.

Harvesting

The seed pods become brown and parchment-like when the seeds are near maturity. The pods do not shatter very readily and it is therefore better to harvest under very dry conditions if a combine is to be used. The dry pods are then relatively brittle and seed extraction is easier. Combines with a roller attachment on some types of bean threshers are most suitable. The rollers are adjusted so as to crack (but not crush) the pods and unthreshed pods are returned to the threshing cylinder.

A standard swather is used if the seed crop is cut before threshing, and the material is left to dry further before passing it through a thresher or combine. The cylinder speed should be reduced to 500–600 rpm.

Seed yield and 1000 grain weight

A seed yield of *c.* 1000 kg per hectare is satisfactory, although yields of 1500–2000 kg per hectare are achieved by some seed producers.

The 1000 grain weight of radish is *c.* 10 g.

The main seed-borne pathogens of radish with common names of the diseases they cause

Pathogens	Common names
Alternaria brassicae (Berk.) Sacc.	Grey leaf spot
Alternaria brassicicola (Schw.) Wilts.	Black leaf spot
Alternaria raphani Groves and Skolko, syn. *A. matthiolae* Neergaard	Leaf spot
Colletotrichum higginsianum Sacc.	Anthracnose, leaf spot
Leptosphaeria maculans (Desm.) Ces. and de Not., syns. *Plenodomus lingam* (Tode ex Fr.) Höhnel *Phoma lingam* (Tode ex Fr.) Desm.	Black leg
Rhizoctonia solani Kühn	Damping off, canker
Xanthomonas vesicatoria (Doidge) Dowson var. *raphani* (White) Star and Burkholder	Bacterial spot

SWEDE, RUTABAGA: *Brassica napus* L.

The cultivated swede originated in Europe; in North America it is commonly known as rutabaga. In addition to its culinary value as a winter root crop it is an important stockfeed crop in Northern Europe. The evolution and development of the modern swedes, which are closely related to swede oil rapes and swede fodder rapes, was outlined by McNaughton (1976a).

Swede cultivar description

Season of maturity, suitability for storage in winter.

Root shape   external root colour: violet-purple, bronze, yellow or white; colour of root crown.

Relative height of 'neck' (tuberized epicotyl base).

Resistance to specific pathogens especially powdery mildew (*Erysiphe cruciferarum*) and clubroot (*Plasmodiophora brassicae*).
The agronomy and techniques used for swede seed production are similar to those for turnips, and the two crops are therefore discussed jointly below.

TURNIP: *Brassica campestris* L. *subsp. rapifera*

Turnips, which originated in Europe, are widely cultivated throughout the world; in the tropics they are generally confined to the higher altitudes. The evolution and development of the turnip was discussed by McNaughton (1976b).

Cultivar description of turnip

Diploid or tetraploid; season of maturity, suitability for early production and protected cropping, suitability for storage.
Root   shape: flat, globe or long globe
          colour: skin and flesh, white, purple or yellow, colour of root crown. The range of turnip root shapes is shown in Fig. 9.4

Leaf   shape, colour, pose and relative size

Resistance to specific pathogens, e.g. clubroot (*Plasmodiophora brassicae* Woron.)

Production of swede and turnip seeds

Soil pH and nutrition

Swedes and turnips require a soil pH between 5.5 and 6.8; soils which have a pH at or below the lower end of this range must receive adequate quantities of liming materials during their preparation prior to sowing.
    The general N : P : K fertilizer requirements applied during seed bed preparation are in the ratio of 1 : 2 : 2 or 1 : 1 : 1; the lower nitrogen ratio is used for turnips unless the soil's nitrogen level is low.
    Both crops are susceptible to boron deficiency which causes 'brown heart' of their roots. Boronated fertilizers are used where soils are known to be relatively low in this micro-nutrient.

Fig. 9.4   Range of turnip root shapes.

Supplementary nitrogen is usually applied as a top dressing in the spring, especially in areas or seasons in which rainfall results in a high rate of leaching. However, supplementary nitrogen dressings must be carefully monitored as excessive nitrogenous fertilizers increase the incidence of lodging when the seed crop is maturing.

Plant production

Although 'seed to seed' and 'root to seed' systems are used for both crops, a 'root to seed' system is generally used for basic seed production. The production of swede 'stecklings' for transplanting was widespread in Northern Europe but has been largely replaced by the seed to seed method.

When direct drilled, 2–4 kg are sown per hectare in rows 50–90 cm apart. The crop is thinned to approximately 4–5 cm between plants in the rows. When a crop is to be transplanted, a sowing rate of 3–4 kg/ha will provide sufficient transplants for 6–10 ha. In areas which are subject to severe winter frosts the overwintering plants are protected with straw or other suitable materials. Both crops are sown in the late summer for seed production in the seed to seed system, but turnips are generally less hardy than swedes and early spring sowings are made in areas where the crop would not be expected to survive the winter.

Topping

Both crops are 'topped' to encourage the development of secondary inflorescences on the main shoots. This reduces the overall height of the crop and the possibility of lodging at a later stage. This removal of the growing points also reduces the seed crop's range of maturity period. The top 10 cm of the terminal shoots are removed when the flowering shoots are between 30 and 40 cm high.

Flowering, pollination and isolation

Swedes and turnips are both biennials and the overwintering plants are vernalized. There is a wide range of vernalization requirements among turnip cultivars and although some of the stubble and stockfeed types will flower in their first year from a spring sowing, other cultivars, e.g. Golden Ball, will not (NIAB).

Both crops are mainly insect pollinated although some wind pollination also occurs. The vegetable cultivars of swede readily cross–pollinate with cultivars of agricultural swedes, swede oil rape, swede fodder rape, kale rape and the turnip group. The culinary turnip cultivars cross–pollinate with agricultural turnips, turnip oil rape, turnip fodder rape and the swede types listed above. The recommended isolation distance for swedes and turnips is 1000 m.

Roguing stages for swedes and turnips

Seed to seed

1. *Early vegetative stage, before swelling of the roots* Check leaf type, colour and relative height.

2. *Start of anthesis* Check that flower colour and flower size are according to type.

Root to seed

1. *Early vegetative stage, before swelling of the roots* Check leaf type, colour and relative height.

2. *When roots are lifted for storage (or re-planting, depending on local winter climate and custom)* Check that root shape, relative size, colour of root and shoulder are according to type.

3. *When re-planted crop reaches anthesis* Check that flower colour and flower size are according to type.

Basic seed production

Only the root to seed system is used for basic seed production. Roots for selection should be grown to normal market maturity by the autumn of their first year. They are then selected on root characters as described above. The early maturing turnip cultivars are sown later than for the early market crop, otherwise the roots are too large by the late summer and autumn and have a tendency to rot during storage.

The selected roots are stored in frost-proof conditions in moist peat or sand and re-planted in the late winter when no further severe frosts are expected. There is an increased tendency to re-plant the selected roots in polythene tunnels at the end of winter.

Harvesting

Both seed crops have a tendency to shatter readily and therefore care in cutting is required before seeds are lost unnecessarily. Generally the best sign that the bulk of the seeds are near maturity is that the plant haulm turns from green-brown to parchment colour. The crop is cut with a swather and left in windrows until the seeds are mature and separate easily from their pods. The material is then either picked up by a combine or fed into a thresher.

Seed yield and 1000 grain weight

A satisfactory yield for both crops is 1500 kg/ha although yields of up to 2500 kg/ha are achieved in the best seed production areas of the USA. The 1000 grain weight of turnip is *c.* 4.3 g and the 1000 grain weight of swede is *c.* 3.3 g.

The main seed-borne pathogens of *Brassica* spp. with common names of the diseases they cause are listed above, at the end of the section on cole crops.

WATERCRESS: *Rorippa nasturtium-aquaticum* L. Hayek syn. *Nasturtium officinale* R.Br.

Origins and types

The early history of watercress, which originated in Northern Europe, was discussed by Howard (1976). It is now cultivated in Europe, Asia, North America, South Africa and Australasia. Watercress is grown commercially as an aquatic crop although plants will thrive in a soil substrate without aquatic conditions. The leaves and shoots are eaten as a salad, or in some areas they are cooked.

The species cultivated widely in Great Britain and to some extent elsewhere in Europe is *Rorippa nasturtium-aquaticum* L. Hayek; this is commonly called 'green-cress'. It has largely replaced *Rorippa microphylla* (commonly called 'brown-cress') which is sterile and was therefore propagated vegetatively. In addition to its more attractive colour the 'green-cress' has other advantages.

It is less susceptible to crook root disease (Spencer and Glasscock 1953) and virus-free stocks can be raised from seed (Tomlinson 1957). There has therefore been a very keen interest in developing seed production techniques for this crop.

There are no cultivars although many growers have produced and maintained their own selections or 'strains'.

Agronomy and crop production

The crop is maintained in the same way as the market crop except that beds for seed production are inspected and rogued especially for off-types and presence of viruses. For further information on how this crop is grown for market the reader is referred to Lyon and Howard (1952) and Stevens (1975).

Flowering and pollination

Work by Bleasdale (1964) showed that 'green-cress' is a long-day plant which flowers in response to the increasing daylength of spring and summer in Northern Europe. As a result of this work it was suggested that late flowering strains could be selected. Although late flowering strains would prolong the market season of the vegetative shoots, the length of season for seed production would be reduced. With this in mind some seed producers have developed techniques of either transferring plants from aquatic beds to a soil substrate in greenhouses or putting temporary plastic structures, such as inflated 'bubble-greenhouses', over the watercress beds.

Watercress flowers can be both cross-pollinated and self-pollinated (Johnson 1974). Several insects are pollinating agents, including Diptera but there are no specific references in the literature to suitability of different species for watercress seed production in confined areas.

Isolation and roguing

There are no specific recommendations in the literature to isolation distances for seed production of this crop, but a minimum distance of 1000 m would be appropriate.

The beds of plants for seed production are generally rogued during the autumn, winter and spring prior to flowering. Plants showing symptoms of virus infection, any other pathogens, pale green leaf colour and early flowering, are discarded. The most important virus of watercress is turnip mosaic virus, although this is not seed-transmitted. Smaller, weaker plants which are otherwise healthy are also discarded in order to maintain or increase the vegetative productivity of a strain.

Harvesting and 1000 grain weight

Watercress, like other Cruciferae seed crops, has a tendency to shatter. The flowering stems are cut and placed in paper lined trays to dry as soon as the majority of seeds are mature. The seeds are a pale yellow colour at this stage and later turn a darker brown as they age. The seeds are usually extracted by hand rubbing the plant material and then passing through a series of seed sieves to remove pieces of plant debris.

The 1000 grain weight of watercress seed is *c.* 2.8 g.

REFERENCES

Anon. (1980) Top French seed starts from stumps, *Grower* **93**(7), 16.

Banga, O. (1976) *Radish.* In *Evolution of Crop Plants*, N.W. Simmonds (ed.). Longman, London and New York.

Becker, G. (1962) Rettich und Radies (*Raphanus sativus*), *Handbuch der Pflanzenzüchtung* **6**, 23–78.

Bleasdale, J. K. A. (1964) The flowering and growth of watercress (*Nasturtium officinale* R.Br.). *J. Hort. Sci.* **39**, 277–83.

Crisp, P. and Lewthwaite, J. J. (1974) Curd grafting as an aid to cauliflower breeding, *Euphytica* **23**, 114–20.

Crisp, P. and Walkey, D. G. A. (1974) The use of aseptic meristem culture in cauliflower breeding, *Euphytica* **23**, 305–13.

Crisp, P., Gray, A. R. and Jewell, P. A. (1975a) Selection against the bracting defect of cauliflower, *Euphytica* **24**, 459–65.

Crisp, P., Jewell, A. and Gray, A. R. (1975b) Improved selection against the purple colour defect of cauliflower, *Euphytica* **24**, 177–80.

Faulkner, G. J. (1962) Blowflies as pollinators of brassica crops, *Commercial Grower* **3457**,807–9.

Faulkner, G. J. (1978) Seed production of F_1 hybrid Brussels sprouts, *Acta Horticulturae* **83**, 37–42.

George, R. A. T. and Evans, D. R. (1981) A classification of winter radish cultivars, *Euphytica* **30**, 483–92.

Howard, H. W. (1976) *Watercress.* In *Evolution of Crop Plants*, N. W. Simmonds (ed.). Longman, London and New York.

Johnson, A. G. (1958) How to increase top quality sprout yields, *Grower*, 1 November.

Johnson, A. G. (1974) Possibilities and problems in the breeding of watercress. *Symposium on Research on the Watercress crop*, pp. 28–32. Bath University.

Kahangi, E. M. and Waithaka, K. (1981) Flowering of cabbage and kale in Kenya as influenced by altitude and GA application. *J. Hort. Sci.* **56**(3), 185–8.

Lyon, A. G. and Howard, H. W. (1952) Science and watercress growing, *Agriculture* **59**, 123–8.

McNaughton, I. H. (1976a) Swedes and rapes. In *Evolution of Crop Plants*, N. W. Simmonds (ed.). Longman, London and New York.

McNaughton, I. H. (1976b) *Turnip and relatives.* In *Evolution of Crop Plants*, N. W. Simmonds (ed.). Longman, London and New York.

NIAB *Growing Turnip for Seed.* Seed growers leaflet No.3. National Institute of Agricultural Botany, Cambridge.

Nieuwhof, M. (1969) *Cole Crops.* Leonard Hill, London.

Spencer, D. M. and Glasscock, H. H. (1953) Crook root of watercress, *Plant Path.* **2**, 19.

Stevens, C.P. (1975) *The Production of Watercress from Seed.* MAFF.

Thompson, K. F. (1976) Cabbages, kales etc. In *Evolution of Crop Plants*, N. W. Simmonds (ed.). Longman, London and New York.

Tomlinson, J. A. (1957) Mosaic diseases of watercress. In *Annual Report, National Vegetable Research Station*, 61.

Walkey, D. G. A., Cooper, V. C. and Crisp, P. (1974) The production of virus-free cauliflower by tissue culture, *J. Hort. Sci.* **49**, 273–5.

Watts, L. E. (1960) The use of a new technique in breeding for solidity in radish, *J. Hort. Sci.*, **35**, 221–6.

Watts, L. E. (1980) *Flower and Vegetable Plant Breeding.* Grower Books, London.

Watts, L. E. and George, R. A. T. (1963) Vegetative propagation of autumn cauliflower, *Euphytica* **12**, 341–5.

Winter, E. J. (1974) *Water, Soil and the Plant.* Macmillan, London.

FURTHER READING

Austin, R. (1966) The growth of watercress *Rorippa nasturtium-aquaticum* (L) Hayek from seed as affected by the phosphorus nutrition of the parent plant, *Plant and Soil*, 24(1), 113–20.

Biddington, N. L. and Ling, B. (1983) The germination of watercress (*Rorippa nasturtium-aquaticum*) seeds. I. The effects of age, storage, temperature, light and hormones on germination, *J. Hort. Sci.* **58**(3), 417–26.

Biddington, N. L., Ling, B. and Dearman, A. S. (1983) The germination of watercress (*Rorippa nasturtium-aquaticum*) seeds. II. The relationship between seed colour and germination, *J. Hort. Sci.* **58**(3), 427–33.

Bowring, J. D. C. and Day, M. J. (1977) Variety maintenance for swedes and kale. *J. Nat. Inst. Agric. Bot.* **14**, 312–20.

Crane, M. B. and Mather, K. (1943) The natural cross-pollination of crop plants with particular reference to the radish, *Ann. Appl. Biol.* **30**(4), 301–8.

Faulkner, G. J. (1974) Factors affecting field-scale production of seed of F_1 hybrid Brussels sprouts, *Ann. Appl. Biol.* **77**,181–90.

Faulkner, G. J. (1978) Seed production of F_1 hybrid Brussels sprouts, *Acta Horticulturae*, **83**, 37–42.

Hodgkin, T. (1981) Some aspects of sib production in F_1 cultivars of *Brassica oleracea, Acta Horticulturae* **111**, 17–24.

Honma, S., Bukovac, M. J. and Heeckt, O. (1961) Studies on the asexual propagation of cauliflower. *Proc. Amer. Soc. Hort. Sci.*, **78**, 343–8.

Ito, H. and Saito, T. (1961) Time and temperature factors for the flower formation in cabbage. *Tohokli J. Agric. Res.* **12**, 297–316.

MAFF (1981) *Insect Pests of Brassica Seed Crops.* Leaflet 576. Ministry of Agriculture, Fisheries and Food.

Nieuwhof, M. (1958) Vegetative maintenance and propagation of cauliflower, *Euphytica* **7**, 170–8.

North, C. (1953) Three methods for vegetative propagation of *Brassica oleracea, J. Roy. Hort. Soc.* **78**, 106–11.

Salter, P. J. (1961) The irrigation of early summer cauliflower in relation to stage of growth, plant spacing and nitrogen level, *J. Hort. Sci.* **36**, 241–53.

Sneddon, J. L. (1962) A note on vegetative propagation of some forms of *Brassica oleracea, J. Nat. Inst. Agric. Bot.* **9**, 145–8.

Stokes, P. and Verkerk, K. (1951) Flower formation in brussels sprouts, *Meded. Landbouwhogesh, Wageningen* **50**, 141–60.

Thomas, T. H. (1980) Flowering of Brussels sprouts in response to low temperature treatment at different stages of growth, *Scientia Horticulturae*, **12**, 221–9.

Watts, L. E. and George, R. A. T. (1957) Classification and performance of varieties of radish. *Eighth Annual Report of National Vegetable Research Station*, pp. 3–15.

Weiring, D. (1964) The use of insects for pollinating brassica crops, *Euphytica* **13**, 24–8.

Wills, A. B. and North, C. (1978) Problems of hybrid seed production, *Acta Horticulturae* **111**, 31–6.

10 CUCURBITACEAE

The genera of this family are widely distributed in the world. They are referred to as 'vine' crops in some areas, especially in North America. All the species are susceptible to frost and while most are important outdoor crops, e.g. watermelon in the southern states of the USA and many other tropical and arid areas, some, e.g. cucumber, have become important protected crops, especially in Northern Europe.

Some of the crops, e.g. watermelon and melon are not regarded as vegetables but as 'fruits' by some authorities because they are eaten as dessert rather than cooked or prepared for culinary dishes. However, because of the production methods of the market crops and their seeds they are generally considered with the vegetable crops.

The majority of Cucurbitaceae species have unisexual flowers borne on monoecious plants, *Telfaria pedata* (*Telfaria* nut or fluted pumkin) has unisexual flowers on separate plants.

There are several interesting problems in the production of cucurbit seed, including the extraction of seed from 'wet' fruit (discussed in detail in Chapter 4), the cross-compatibility of some different crops or cultivar types within the same genera, and the positive identification of plants for seed production based on immature fruit characters. This last point, coupled with the fact that the majority of cultivated cucurbits are highly cross-fertilized, leads to the need for careful attention to isolation and roguing especially for the production of basic seed.

The main cucurbit crops dealt with in this chapter include:

Citrullus lanatus (Thunb.) Mansf. syn. *C. vulgaris* Schrad.	watermelon, egusi
Cucumis melo L.	cantaloupe, melon, sweet melon
Cucumis sativus L.	cucumber
Cucurbita maxima Duch. ex Lam.	pumpkin, winter squash, chinese pumpkin, crookneck squash, squash, gourd
Cucurbita moschata (Duch. ex Lam.) Duch. ex Poir	pumpkin, winter squash
Cucurbita mixta Pang.	pumpkin, winter squash
Cucurbita pepo L.	marrow, vegetable marrow, courgette

WATERMELON: *Citrullus lanatus* (Thunb.) Mansf.

This crop is widely grown throughout the tropics, sub-tropics and arid regions of the world. It originates from Africa but there are references to the very early cultivation of the species elsewhere in the world (Hedrick 1972). In some parts of Africa local cultivars with relatively bitter fruits are grown for their seeds which are roasted and eaten; these are referred to as 'egusi', but production for this purpose is not generally included as part of seed production.

Watermelons have been further selected and improved by seedsmen and plant breeders, especially in the USA, where a very wide range of cultivars is maintained. Some of these cultivars, e.g. cv. Charleston Gray, originally bred in the USA, are cultivated in many areas of the world.

The watermelon, a vigorous annual which covers a large area of ground with its sprawling stems, can survive relatively dry conditions because it roots deeply. For this reason it has become established as an important crop in many developing countries, especially where arid conditions prevail.

Cultivar description of watermelon

Season

Use

Foliage cover

Vine vigour

Fruit
   External shape round, oval or long
            size, relative weight
            rind colour, bi-colour, striped (Fig. 10.1)
  Internal thickness of rind
            flesh colour
            intensity of flesh colour, especially towards the centre of the fruit
            absence of central cavity

Seed characters   colour, striped, relative size

Resistance to disease, especially *Fusarium* sp., Anthracnose (black end to fruit) and sunscald.

Nutrition

Many small-scale watermelon producers apply a bulky organic manure during soil preparations at a rate of up to 25 tonnes per hectare if locally available, but this is often not possible for large-scale production.

Fig. 10.1   Examples of the colouring and rind patterns of watermelon cultivars.

In the absence of organic manures the base dressing ratio applied should be 1 : 1 : 1 N :P : K according to the nutrient status of the soil. Higher nitrogen ratios than this should be avoided otherwise the plants make excessive vegetative growth.

Agronomy

In most areas of the world where watermelons are grown for seed production, seeds are sown direct into the field in preference to propagation in nurseries and then transplanting.

The plants are either grown on 'mounds', flat ridges or on the flat. The system adopted locally depends on the irrigation system to be used and to some extent on custom. Mounds or ridges are used in conjunction with a furrow or similar irrigation system. The seeds are spot sown, usually 2 to 3 seeds per station 90–120 cm apart in the row with 120–180 cm between the rows; 1 to 3 kg of seed are sufficient to sow 1 ha. A higher seeding rate is required when the seeds are drilled.

Seed is slow to germinate early in the season if soil temperatures are low. The plants are very frost sensitive and local custom generally avoids a sowing date before the incidence of frosts has passed.

The seedlings are thinned out to their final stand when the first true leaves are showing. Care must be taken with crops scheduled for seed production to ensure that only one seedling is left per station.

The frequency of irrigation will depend on soil type and climate, but because watermelon plants develop a deep and extensive root system applications can be kept to a minimum, but in dry regions sufficient irrigation should be applied prior to sowing to restore the soil to field capacity.

Flowering

Watermelon is day-neutral and there are therefore no problems in flower initiation. However, plant and fruit development are poor when ambient temperatures are less than 25 °C.

Pollination

Watermelon flowers are insect pollinated, mainly by honeybees. The plants are self-compatible, but because the flowers are unisexual a high percentage of cross-pollination occurs.

It is normal practice in the USA to place colonies of hive bees on the perimeter of watermelon fields (Fig. 2.1). This is done to increase seed yield and it is claimed also that by supplying a high population of pollinating insects adjacent to the site of seed production, the incidence of cross-pollination with other fields of watermelon is minimized.

Isolation

The minimum recommended isolation distance of watermelons for seed production is 1000 m. Recommended distances for stock seed are at least 1500 m. It is important to ensure that seed crops are also isolated from market crops.

Harvesting

Sufficient time must be allowed for the seed to reach maturity which is at least a week later than the optimum stage for marketing the fruit. The harvesting stage for seed production can be confirmed when the tendrils have withered on the shoot bearing the fruit. Another sign that the fruit has reached maturity is the colour change from a green/white to pale yellow of the underside of the fruit (i.e. the surface which has been resting on the soil).

The method of collecting fruit for seed extraction depends on the scale of operation. In the USA, where fields for production of watermelon seed are often at least 10–20 ha the whole operation is mechanized. In countries which use a high percentage of hand labour, the entire operation may be done by hand, especially in smaller areas for stock seed or commercial seed production.

The specialist machines which have been developed in the USA are either self-propelled or trailed by a tractor. The self-propelled machines are capable of picking up the fruits from the plants, but this is done when the entire crop has reached maturity as it is a once over harvest which collects all the fruit and destroys the plants (Fig. 10.13).

Hand-harvesting methods are based on the selective cutting of mature fruits which are put direct into a cucurbit seed extractor. This machine is towed through the field at a speed compatible with the rate of cutting, according to the number of workers available.

An alternative system is to cut the mature fruits and place them in wind-rows or heaps either to await the seed extractor which goes through the field later, or they are collected for immediate transportation to a central area where the extraction is done.

Seed extraction

The seeds in the watermelon fruit, unlike most cucurbits, are distributed throughout the central area of the fruit pulp (Fig. 10.2) and not in a central fruit cavity. Therefore hand extraction methods depend on fruit maceration rather than a scooping process.

The macerated pulp containing the seeds is washed by running water into a screen; this operation separates the pieces of rind and coarse material from the seeds and fine pulp. The seeds pass through to a finer meshed screen which retains them. The better the initial maceration of pulp, the more efficient is the separation by screens, resulting in a cleaner seed sample.

Fig. 10.2   Transverse section of watermelons; left 'seedless' type with small number of undeveloped seeds; right fertile black seeded cultivar.

Fermentation is not normally used in the extraction or cleaning of watermelon seed because the seed easily discolours and potential germination is reduced by fermentation.

Drying

The drying of watermelon seed must commence as soon as the extraction is completed. Large rotary driers (Fig. 10.3) are used by specialist watermelon seed producers especially for a preliminary drying period. The control of air temperature is not very accurate with rotary driers and many seed producers prefer the batch driers with a rotary paddle (Fig. 4.10). The initial air temperature for drying is usually between 38 and 41 °C and as the pieces of fruit and rind debris dry off (i.e. there is no obvious free moisture when a piece of pulp is pressed between the fingers) the temperature is reduced to 32–35 °C. The drying is continued until the seed moisture content does not exceed 10 per cent. If seeds are going to be stored in sealed vapour-proof containers they are dried down to a moisture content of 6 per cent.

Seed yield

The average seed yield of most watermelon cultivars under good conditions

Fig. 10.3   Rotary drier for preliminary drying of large seeded cucurbits.

is 400 kg/ha. The popular cultivar 'Charleston Gray' is a relatively low seed yielder and an average yield is only 250 kg/ha. Under the relatively poor cultural conditions found in some tropical countries, seed yields of the better yielding cultivars are as low as 100 kg/ha with cv. Charleston Gray yielding even less. A relatively short growing season also affects seed yield adversely. A lower yield of fruit is achieved when the fruit is left on the plant for seed production than if fruits are promptly harvested as they reach market maturity.

The 1000 grain weight depends on the cultivar but the approximate weight of 1000 seeds is 113 g.

Watermelon roguing stages

1. *Before flowering*: check vegetative characters.
2. *At early flowering*: check trueness to type of developing fruit.
3. *Fruit developing*: as for stage 2.
4. *Marketable fruit*: check fruit characters.

Stock seed production

Foundation seed is maintained by selfing single plant selections which are grown in isolation. After harvesting the seed separately from each selected plant, the individual seed-lots are progeny tested and the approved material is bulked for further multiplication.

The main seed-borne watermelon pathogens with the common names of the diseases they cause

Pathogens	Common names
Colletotrichum lagenarium (Pass.) Ell. and Halst., syn. *C. orbiculare* (Berk. and Mont.) Arx.	Anthracnose
Didymella bryoniae (Auersw.) Rehm. syns. *Mycosphaerella melonis* (Pass.) Chiu and Walker. and *Phyllostica citrullina* Chester	Gummy stem blight, black rot
Fusarium oxysporum Schlecht ex f.sp. *niveum* (E.F.Sm.) Snyder and Hansen syn. *F. citrulli* Taub.	Wilt
Pythium aphanidermatum (Edson) Fitzp.	Seed rot
Pseudomonas sp.	
Virus	Squash mosaic virus

CUCUMBER: *Cucumis sativus* L.

This species is widely cultivated for the production of its edible fruits which are used as a salad. Small fruited types of cucumbers are also used in pickles. In the Far East the young leaves are eaten raw or cooked similarly to spinach. Cucumbers are cultivated as a field crop in most areas of the world under frost-free conditions, but are also an important greenhouse crop in Northern Europe. The traditional greenhouse cultivars develop parthenocarpic fruits if not pollinated and this character is now incorporated into many of the out-door types.

Origins and types

There is a wide range of cucumber fruit types showing a diversity of fruit size, shape, colour and spininess. An indication of the diversity can be seen from Fig. 10.4.

Fig. 10.4   Range of cucumber fruit types.

Cucumber cultivars are generally classified into four main groups:

1. Field cucumbers, with prominent black or white spines.
2. The greenhouse or forcing type (often referred to as 'English cucumber'). These types have relatively long and spineless fruits which can be produced parthenocarpically.

3. The Sikkim cultivars (originating from India) with reddish-orange fruits.
4. The small fruited cultivars frequently used as 'gherkins' for pickling.

There is a further group found in several diverse areas of the world including the USA and the Far East which includes the 'apple' cucumbers, which have ovoid- to spherical-shaped fruits.

Cultivar description of cucumber

Use	salad, pickling; greenhouse or outdoor production
Season	earliness, suitability for protected cropping
Fruit characters	relative size and shape
	shape of basal end
	rind colour when ready for market
	rind colour when mature for seed production
	spines: degree of spininess, spine colour (black or white) and relative size of 'warts' at bases of spines
	ability of fruit to retain characteristic shape if picking delayed
	freedom from bitterness

Vegetative characters: plant habit, tendency to produce lateral shoots. Resistance to specific pests and pathogens, e.g. nematodes

Nutrition

Cucumber plants require a soil with a pH of 6.5 or slightly above. The crop responds to soils with a relatively high organic matter content, therefore, if possible, the soil should receive a dressing of up to 80 tonnes per hectare of decomposed organic manure during the early stages of preparation.

A suitable nutrient ratio application of N : P : K 1 : 2 : 2 should be applied during the final stages of seed bed preparation, but the nutrient value of any bulky organic manures already applied should be taken into account. The nitrogen is increased on soils with a high phosphorus and potassium status. A higher proportion of nitrogen is also necessary where frequent irrigation is required in order to allow for leaching; in this situation the N : P : K ratio should be nearer to 2 : 1 : 1, with approximately half the nitrogen applied as a top dressing about a month after seedling emergence. Care must be taken to avoid foliar scorch from this operation.

Plant establishment

Seeds of the open-pollinated cultivars are normally sown direct into the field at stations 10–12 cm apart, with up to 2 m between the rows. In arid areas where irrigation channels are necessary, the seeds are sown on flat ridges which are up to 30 cm high. The seedlings are thinned to a single plant per station soon after they have emerged. Similar systems are used for hybrid seed production except that a closer row spacing of 50 cm is usually adopted and there are six to eight rows of the female parent between two male parent (or pollinator) lines; this pattern is repeated across the field.

The plants are 'stopped' by pinching out the initial leader between three to five leaves; two main laterals are subsequently secured. About four or five fertilized fruits are obtained per plant in the female rows. A similar stopping system is usually adopted for the open-pollinated cultivars, see Fig. 10.5.

Fig. 10.5   Cucumber plant which has been 'stopped' to limit lateral growth and number of fruits to four.

Sex expression in cucumber flowers

Cucumber plants produce both male and female flowers. The monoecious cultivars normally produce these in approximately equal proportions, although there is a tendency to produce only male flowers initially.

Sex expression in cucumber is generally influenced by environment. Under long days and high light intensities male (staminate) flowers predominate, whereas under short days and low light intensities female (pistillate) flowers predominate.

Recent breeding and selection work with cucumbers has resulted in the production of cultivars which under normal conditions bear predominantly or completely female flowers. These are referred to as gynaecious cultivars. The advantages of these, many of which are F_1 hybrids, are that they are earlier, higher yielding and all fruits are parthenocarpic and therefore seedless. In addition, some of the F_1 hybrids have resistance to specific pathogens.

There is an obvious technical problem for the seed producer who has to manipulate the plant in order to produce a higher proportion of male flowers to ensure an adequate supply of pollen. Different breeders and seed producers have developed their own techniques for producing staminate flowers on cucumbers based on experiences with individual cultivars or lines in specific environments or locations. These are based on the use of gibberellic acid or silver nitrate. Three alternative methods have been developed, these are:

1. Three applications of GA_3 at 1000 ppm, sprayed on at fortnightly intervals, commencing when the plants have two leaves.
2. As above, but using $GA_4/_7$ at 50 ppm.
3. A single application of silver nitrate solution (600 mg/l) before the first flowers open.

Hybrid seed is collected only from the female parent plants; the presence of any pistillate flowers on the male plants is not a problem but it is important that staminate flowers are completely suppressed on the female parent. This is normally achieved by two applications of ethrel (250 ppm), the first when the plants show their first true leaf and the second at the fifth true leaf stage. A visual check must also be made and any male flowers on the female parents removed by hand. The development of suitable male sterile lines or application of a satisfactory gametocide would obviously be a useful development in hybrid cucumber seed production.

A further safeguard which avoids admixture of seeds from the male parents is to rotavate the male parent rows before harvesting commences. This is normal practice in the USA especially when the entire harvesting process is mechanized.

Pollination

Cucumbers are self-compatible but are predominantly cross-pollinated. Pollination is mainly done by bees when the plants are grown as a field crop. *Cucumis sativus* is not cross-compatible with *C. melo*, gourds, marrows, squash, pumpkins or watermelons.

Isolation

Cucumber seed production in fields or plots should be at least 1000 m from all other cucumber crops including those for market production. It is especially important that the different types of cucumbers have sufficient isolation during production of commercial seed. For basic seed production the isolation distance should be at least 1500 m. These recommended distances can be reduced when seed is produced in insect-proof greenhouses.

Cucumber roguing stages and main characters to be observed

Seed crops for commercial category seeds are usually only examined and rogued at stages 3 and 4.

1. *Before first flowers open* Desirable characters – growth habit, vigour and foliage typical of the cultivar.

2. *Early flowering* General plant habit and foliage characters as checked in stage 1. Observable characters of undeveloped fruit, especially colour of spines. Observe if any specific seed-borne disease present.

3. *Fruit setting* As for stage 2. Also:
(a) Satisfactory level of productivity.
(b) Time of main production, e.g. earliness.
(c) Fruit characters, including size, shape and colour.

4. *Ripe Fruit* Colour of ripe fruit in accordance with cultivar description, e.g. fruits either green, yellow, white or orange.

Harvesting

The fruit must remain on the plants until fully mature. This is indicated externally by the development of the ripe rind colour characteristic for the cultivar; additionally the fruit stalk adjacent to the fruit withers when the seed is mature.

In order to confirm the external signs that the seeds are actually mature, several fruits should be cut open longitudinally and the seeds examined. Mature seeds separate easily from the interior flesh.

The mature fruit are hand picked and put into a fruit crusher and seed extractor (as described above for watermelon). Large-scale specialist producers use mechanized harvesting machines incorporating crushers and seed extractors. If the seeds are extracted by hand, the ripe fruits are cut in halves longitudinally and the seeds are scraped out into a container.

The seed and juice mixture can be fermented for about a day before screening and washing the seed in suitable-sized sieves. Water troughs with rifles, as described for tomato seed washing, are used in large-scale operations.

The seed is then dried as described for watermelon. After drying, the seeds are screened to remove any remaining fruit debris. Aspirated screens will remove light and immature seeds.

Seed yield

The average seed yield under field conditions is 400 kg per hectare, although yields of up to 700 kg are often reported. The yield from a single fruit depends on the cultivar and the amount of successful pollination, but estimations can be based on approximately 500 seeds per fruit. The seed yield for

F_1 hybrids with a male : female plant population ratio of 1 : 4 is 300–350 kg per hectare.

Seed weight

The 1000 grain weight of the smaller fruited cucumbers is 25 g. The seeds of the most long fruited greenhouse cultivars have a 1000 grain weight of 33 g.

The main seed-borne cucumber pathogens with the common names of the diseases they cause

Pathogens	Common names
Alternaria cucumerina (Ell. and Ev.) Elliott syn. *A. brassicae* (Berk.) Sacc. f. *nigrescens* Peglion	Leaf spot
Cladosporium cucumerinum Ell. and Arth.	Scab, gummosis
Colletotrichum lagenarium (Pass.) Ell. and Halst.	Anthracnose
Corynespora cassiicola (Berk. and Curt.) Wei syns. *Cercospora melonis* Cooke and	
Helminthosporium cassiicola Berk and Curt.	*Cercospora* leaf spot
Didymella bryoniae (Auersw.) Rehm, syns. *Mycosphaerella melonis* (Pass.) Chiu and Walker	
M. citrullina (C. O. Smith) Grossenb and *Phyllosticta citrullina* Chester	Leaf spot, black rot
Fusarium oxysporum Schlecht. ex. Fr.	Fusarium wilt

Pseudomonas lachrymans (E. F. Smith and Bryan)
    Carsner
	Cucumber angular leaf
	    spot

Viruses	Cucumber green mottle
	    mosac virus
	    syn. *Cucumis* virus 2
	 Cucumber mosaic
	    virus
	    syn. cucumber mosaic
	    virus 1 (mosaic virus)

MELON: *Cucumis melo* L.

The fruits of this species are eaten as a dessert and are not normally cooked although some are used in preserves. There is a very wide diversity of fruit types and local selection in different areas of the world has led to a large number of cultivars. Purseglove (1968) divided the cultivated melons into four types.

1. The European cantaloupe melon, with thick, scaly, rough rind which is sometimes grooved.
2. The musk-melon, cultivated in the USA, with smaller fruits than the cantaloupe melon, and the rind finely netted or almost smooth with shallow ribs.
3. The Casaba or winter melon with large fruits, late maturing, with good storage quality. The yellow or greenish rind is usually smooth, some cultivars striped or splashed; the flesh has a slightly musky flavour.
4. The oriental types, with elongate fruits nearer to cucumbers in shape and often used as vegetables.

Cross-pollination between different cultivars occurs freely and there are therefore many intermediate types.

Melons are cultivated throughout the tropics, sub-tropics and to a limited extent in parts of the temperate regions which have relatively long warm summers. Melons have also been grown as protected crops in heated greenhouses in Northern Europe, but their production as protected crops is now largely confined to unheated polythene tunnels in the Mediterranean region where they have become an important export crop.

A range of modern cultivars is shown in Fig. 10.6 and transverse sections of the same fruits are shown in Fig. 10.7.

Cultivar description of melon

Season   outdoor type, suitability for greenhouse production, suitability for shipping, storage

Fig. 10.6   Examples of the range of external characters found in modern melon cultivars.

Fig. 10.7   Transverse sections of the melons shown in Fig. 10.6 showing internal
fruit characters.

Fruit characters   Shape: ratio of length to width
                      rind: external colour, external texture
                      flesh: colour, relative sweetness, relative width of seed
                      cavity and flesh (Fig. 10.7)

Specific pest and pathogen resistance

Nutrition

All types of melons require soils with a pH of between 6.0 and 6.8.

The crop has traditionally been given bulky organic manures incorporated
during preparation of the fields although many areas of the world produce
melons of excellent quality and high seed yields without the application of
bulky organics. This is very evident in the Middle East and parts of Asia
where melons are grown in reclaimed desert soils with available irrigation.
The traditional applications of bulky organic manures for this crop can be
replaced by using rotations of forage crops, such as alfalfa which improve
soil structure.

Fertilizers applied during the final stages of seed bed preparation should
contain a N : P : K ratio of 1 : 1.5 : 2. Although appropriate reductions of
the phosphorus and potassium levels must be made to allow for residues
which are present. Where leaching is likely to be heavy some of the nitrogen
can be applied as a top dressing when flowering commences, although later
applications of nitrogen should be avoided as this can delay fruit maturity and
ripening.

Sowing and Spacing

Melons are grown either on the flat, ridges or raised beds. The choice of system usually follows the local custom which has developed according to irrigation systems available and the soil's drainage character.

Seeds are usually sown direct in twos or threes at stations, thinning to a single plant per station as soon after seedling emergence as practical. The row width used depends on the vigour of the cultivar as well as on the need for irrigation channels. Rows are from 1.25 to 2.0 m apart. Distances between plants within the rows are 90 cm at the narrower row widths and down to 30 cm at the wider row widths; 2 kg of seed will sow 1 ha. Large-scale producers drill the seed on the flat.

The leaders and laterals are stopped when there are about four fruits per plant.

Pollination

Supplementary hives should be placed adjacent to fields where the natural bee population is relatively low. Plants can be hand pollinated when small numbers of plants are grown for stock seed production, especially in greenhouses, but this is not normally economic for large-scale production.

Isolation

The ideal minimum isolation distance for commercial seed production is 1000 m although some authorities specify distances of only 500 m. It is extremely important that there is adequate isolation between commercial crop and seed crop fields of the different types of melons as they are very cross-compatible.

Melon roguing stages and main characters to be observed:

1. *Before flowering*: vegetative habit, characters of undeveloped fruit.
2. *After flowering commences*: check fruit type.
3. *Fruit set*: check fruit type and colouring.
4. *Harvesting*: check fruit type and colouring.

Resistance to downy mildew (*Pseudoperonospora* spp.), powdery mildew (*Sphaerotheca* sp.) and bacterial soft rot (*Erwinia* sp.).

Harvesting

Cantaloupe and musk-melons tend to separate from the stem at the base of the fruit as the fruit becomes fully mature. This stage of separation by for-

Fig. 10.8   Melons which are almost mature showing position of abscission layer
formed at 'full slip'.

mation of an abscission layer is usually referred to by melon growers as 'full
slip' (Fig. 10.8). Large-scale melon seed producers leave the entire crop of
fruit to separate in this way before passing through the field with a cucurbit
seed harvester or picking up by hand and conveying the fruit in baskets to
the seed extractor which moves through the field.

The winter melons do not form an abscission layer or 'full slip' when
mature; maturity is indicated by colour change from green to yellow or yellow to white (according to the rind colour of cultivar). In addition to the
external colour change the blossom end of the fruit softens and the coat becomes waxy and its aroma increases.

Melon seed is not fermented before washing to separate the seed from other
fruit material. After washing the seed is dried as described for watermelon.
Final separation of dried fruit debris and light seeds can be made by passing
the seed-lot through an aspirated screen cleaner.

Seed yield

The average yield of seed is about 300 kg per hectare, although the best
melon seed production areas can achieve up to 600 kg.

Seed weight

The 1000 grain weight of melon seed is 25 g.

The main seed-borne melon pathogens with common names of the diseases they cause

Pathogens	Common names
Cladosporium cucumerinum Ell. and Arth.	Scab
Colletototrichum lagenarium (Pass.) Ell. and Halst.	Anthracnose
Fusarium sp.	Wilt
Pleospora herbarum (Pers. ex Fr.) Rabenh., syn. *Stemphylium botryosum* Wallr.	Leaf spot
Viruses	Cucumber mosaic virus (syn. Cucumis virus)
	Melon mosaic virus
	Musk-melon mosaic virus (syn. Marmor melonis)
	Squash mosaic virus
	Tobacco ringspot virus

SQUASHES, PUMPKINS AND MARROWS: *Cucurbita* spp.

Summer squash	*Cucurbita pepo* L.
Winter squash	*C. maxima* Duch. ex Lam.
	C. mixta Pang.
	C. moschata (Duch. ex Lam.) Duch. ex Poir
	C. pepo L.
Pumpkins	*C. maxima* Duch. ex Lam.
	C. mixta Pang
	C. moschata (Duch. ex Lam.) Duch. ex Poir
Cushaw	*C. mixta* Pang
Vegetable marrow	*C. pepo* L.

Origins and types

The *Cucurbita* spp. originated in the arid areas of Central America. The species are now widely grown in the tropics and arid regions of the world and many local selections and cultivars exist. They are less important in the humid tropics but are cultivated as summer crops in the temperate areas of the world. Suitable cultivars of *C. pepo* are cultivated as greenhouse crops in Northern Europe in addition to the production as field crops.

According to Purseglove (1968) there is considerable confusion as to the identity of the cultivated species within this group but his classification provides a very useful key to the types. The same common names are frequently used for different species and the American and English common names such as squash, pumpkin and marrow are applied indiscriminately to their fruits.

Because of the similarity in their production methods, the genera listed above are considered here in one group.

Cultivar description of squash, pumpkin and marrow

Open pollinated or F_1 hybrid
Season
Use   e.g. immature fruit, mature fruit or storage

Vegetative habit   trailing (vine) or bush, leaf characters.

Fruit (Figs. 10.9 and 10.10).

  External
                              shape
                              size
                              rind colour, bi-colour, striped
                              rind texture, prominant ribs.

  Internal
                              flesh colour
                              shape and relative size of central cavity

Resistance to specific pathogens, e.g. *Fusarium* sp.

Fig. 10.9   Range of fruit shapes and colouring in 'summer squash' or 'marrow'

Fig. 10.10   Range of fruit types in 'pumpkins'.

Nutrition

This group of *Cucurbita* spp. are moderately tolerant of acid conditions and are produced successfully on soils with a pH between 5.5 and 6.8.

While these vegetables respond to applications of up to 30 tonnes per hectare of bulky organic manures during site preparation, reasonable crops can be produced by appropriate applications of inorganic fertilizers. The optimum N : P : K ratio is 1 : 2 : 2 with appropriate deductions made for nutrients applied as bulky organic manures. Most cucurbit seed producers apply a top dressing of 1 : 1 : 1 at a rate of 60 kg per hectare when the first fruits start to set. On soils with a high P and K status top dressings of only nitrogen are used. The nitrogen top dressing is particularly important when leaching occurs.

Sowing

There are three methods of field production: growing on the flat, on flat ridges or on mounds or 'hills'. The system adopted depends on the irrigation system and efficiency of soil drainage as cucurbits are not successful on waterlogged soils. The preparation of mound or 'hill' systems is very labour intensive as it is difficult to mechanize the operation.

The ultimate plant density depends on the type of cucurbit and whether it is trailing (vine) or bush in its mode of vegetative growth.

Row spacings of 90 cm to 3.5 m are used according to the vigour of the type to be grown. The lower row width is used for less vigorous and bush types whereas the wider row widths are used for the vigorous trailing cultivars. The distance between plants within the rows is usually the same as the distance adopted between the rows, although where production of the crop is fully mechanized a closer spacing within the rows is used to compensate for a wider row spacing and allows for machinery to be used during cultivations. The sowing rate is between 2 and 4 kg per hectare depending on the plant density required. It is normal practice to sow about three seeds per station and single the seedlings after emergence.

Pollination

Each of the *Cucurbita* spp. discussed here is cross-pollinated within the species. While there is some cross-incompatibility between some pairs of these species, for seed production purposes it is safer to assume that crossing between types will take place. The reason for this assumption is that it is not always possible to confirm to which of the four species a particular cultivar belongs.

Pollination is generally by bees although some other insect species are known to pollinate *Cucurbita* flowers.

Isolation

The recommended isolation distance between seed crops within this group is 1000 m. It is also important to avoid areas where even a few of these types are grown domestically for fruit production.

Crops for basic seed production should be isolated by at least 1500 m.

Production of hybrid *Cucurbita* seed

Some seed companies have developed F_1 hybrid cultivars. These have become particularly important in the USA and seed is also produced there for export to other parts of the world.

A ratio of one male parent row to five female parent rows is generally used for hybrid seed production. An ethrel solution (250 ppm) is used to suppress the male flowers on the female rows. It is applied as a spray at three stages of plant development – first true leaf, third true leaf and fifth true leaf. By the time that the first flowers open at about the sixth to seventh leaf stage, all male flowers have been suppressed. Figure 10.11 shows the undeveloped male flowers following application of this spray regime.

By the time that two to three fertile fruits have developed, the ethrel effect has gone. Further sprays would not be effective, so development of later male flowers is stopped by cutting off the growing point with a knife.

Fig. 10.11 *Cucurbita* plant showing suppressed male flowers and position where growing point was removed.

Roguing stages

1. *Early vegetative stage* Check that vegetative characters (e.g. bush or trailing type), foliage and vigour are in accordance with cultivar. Resistance to specific pathogens according to the cultivar description.

2. *Before first flowers open* As above, and check that undeveloped fruit characters on female flower buds are true to type.

3. *First female flowers setting* Check that developing fruits are true to type, and as above.

4. *Fruit developing* Fruit characters true to type, and resistance to specific pathogens according to the cultivar description.

Harvesting

All squashes, pumpkins and marrows take approximately sixteen weeks from anthesis to seed maturity. By this stage the rind has hardened and changed colour. The green types change to a yellow-orange colour and the yellow-golden coloured types change to a straw colour. A crop of 'crook-necked' squash ready for harvest is shown in Fig. 10.12.

For large-scale production fruit may be placed in windrows ready for the mobile thresher and extractor.

Fig. 10.12 'Crook-necked' squash ready for harvest.

Fig. 10.13 'Vine harvester' for harvesting and extraction of cucurbit seeds.

Alternatively they are left on the vine for an automatic vine harvester to pass through the field. A vine harvester capable of picking up, threshing and extracting the seed of a cucurbit crop is shown in Fig. 10.13.

The fruits of some cucurbits such as winter squashes are relatively dry by the time the seed is mature and ready for extraction. In this case water has to be introduced into the separator to wash the seeds free from fruit debris.

After extraction the seeds are washed in troughs and dried (as described for watermelon). *Cucurbita* seeds are not fermented during the cleaning process as this tends to discolour them and reduce the potential germination.

After drying, seeds are passed through an aspirated screen cleaner to remove pieces of dried fruit debris and any light seeds.

Seed yield

The average seed yield is about 500 kg per hectare, but under good pollination and cultural conditions up to 1000 kg per hectare can be obtained.

Seed weight

The 1000 grain weight is approximately 200 g, depending on the cultivar.

The main seed-borne pumpkin, marrow and squash pathogens with common names of the diseases they cause

Pathogens	Common names
Alternaria spp.	Leaf and stem spot
Cladosporium cucumerinum Ell. and Arth.	Scab
Fusarium solani (Mart.) Sacc.f.sp. *Cucurbitae* Snyder and Hansen	*Fusarium* foot rot
Sclerotinia sclerotiorum (Lib.) de Bary	Watery soft rot
Stemphylium spp.	Leaf and stem spot
Xanthomonas cucurbitae (Bryan) Dowson	Bacterial leaf spot
Viruses	Cucumber mosaic virus (syn. *Cucumis* virus 1)
	Musk-melon mosaic virus
	Prunus necrotic ringspot virus (syn. peach ringspot virus).

MISCELLANEOUS CUCURBIT SPECIES

The four cucurbit crops discussed above form the major part of all cultivated Cucurbitaceae. However, in some areas of the world, especially in the tropics, there are many other members of Cucurbitaceae which are of considerable local importance. Those which are more commonly encountered are listed below with their local common names. Annotated information relating to their seed production is included.

Benincasa hispida (Thunb.) Cogn. (ash pumpkin, hairy wax gourd, tallow gourd, wax or white gourd)
This species is important in parts of Asia especially the tropics. Plants for seed production should be selected on vine and fruit characters. Fruits for seeding must remain on the plant until fully mature.

Cucumis anguria L. (West Indian gherkin)
Locally important in South America, especially Brazil, and the West Indies. Plants for seed production are selected as early as possible according to fruit types before any significant cross–pollination can occur. Rogue again when the mature fruits display the important characters.

Lagenaria siceraria (Molina) Standl. (bottle gourd, calabash gourd, white-flowered gourd, zucca melon)
The fruits from a few cultivars of this species are used in culinary preparations, but the majority are grown for their hard external skins or rinds which are used for a wide range of domestic utensils. Plants for seed production should be selected on vine, foliage and fruit characters.

Luffa acutangula (L.) Roxb. (angled gourd, silky gourd, angled loofah, ridged gourd, vegetable gourd)
The young fruits of this species are used as a vegetable in Asia and the Far East. Selection for seed production is based on vine and fruit characters.

Luffa cylindrica (L.) M. J. Roem (dishcloth gourd, smooth loofah, sponge gourd, vegetable sponge)
The main use of this species is as 'vegetable sponge' but the immature fruits of some cultivars are used raw as a salad or cooked in stews and soups. Selection is on vine and fruit characters.

Momordica charantia L. (balsam pear, bitter cucumber, bitter gourd, bitter melon)
This species is a locally important vegetable in India and the Far East. The fruits are pickled or used in curries and in some areas the young vine or leaves are used as a cooked green vegetable. Plants selected for seed production should have fruit characters according to the criteria of local producers. Seed crops should be isolated from fruit crops for market by 100 m.

Sechium edule (Jacq.) Swartz (choyote)
This is a perennial species whose fruits and tuberous roots are used as a cooked vegetable. Replacement stocks and selections to maintain and im-

prove fruit quality are made from plants with optimum fruit and vegetative characters. Each fruit bears a single seed.

Trichosanthes cucumerina L. (snake gourd)
The fruit of this crop is widely popular in Asia and the Far East and locally in Africa. Plants for seed production should be selected for relatively straight and perfect fruit.

REFERENCES

Hedrick, U. P. (ed.) (1972) *Sturtevant's Edible Plants of the World.* Dover publications, New York.
Purseglove, J. W. (1968) *Tropical Crops, Dicotyledons.* Longman, London.

FURTHER READING

Adlerz, W. C. (1966) Honey bee visit numbers and watermelon pollination, *J. Econ. Ent.* **59**, 28–30.
Cobley, L. S. and Steele, W. M. (1976) *An Introduction to the Botany of Tropical Crops.* Longman, London.
Peterson, C. E. and Weigle, J. L. (1958) A new method for producing hybrid cucumber seed, *Mich. Agric. Exp. Sta. Quart. Bull.* **40**, 960–5.
Whitaker, T. W. and Bohn, G. W. (1950) The taxonomy, genetics, production and uses of the cultivated species of *Cucurbita, Econ. Bot.* **4**, 52–81.
Whitaker, T. W. and Davis, G. N. (1962) *Cucurbits.* Leonard Hill, London.

11 LEGUMINOSAE

This family contains a large number of genera grown for food and fodder. Many of the species are also grown for the production of their dried seeds which are frequently referred to as pulses or grain legumes; they include *Pisum sativum* L. (peas), *Phaseolus vulgaris* L. (beans) and *Vicia faba* L. (broad beans). Some, for example *Arachis hypogea* L. (groundnut) and *Glycine max* (L.) Merr. (soybean), are also important crops for processing as well as for fresh vegetable production. These crops are usually considered to be vegetables when grown for the production of their immature seeds or green pods containing immature seeds. The division of species into vegetables and pulses is not always possible as there can be considerable overlap. The classification therefore depends on the method of production and use of end-product. Many aspects of the tropical pulses, including their nutritional value, origins, taxonomy, agronomy and improvement have been dealt with by Smartt (1976).

The vegetables dealt with in this chapter are:

Pisum sativum L.	pea; garden pea
Phaseolus vulgaris L.	dwarf bean, French bean, green bean, snap bean
Phaseolus coccineus L.	runner bean, scarlet runner bean
Vicia faba L.	broad bean

A list of the important tropical vegetable legumes not included above is given in Table 11.1 together with the sowing rates, mean seed yields and 1000 grain weights for each species.

The diversity of the genera cultivated as vegetables in the tropics and sub-tropics is outlined by Grubben (1977).

PEAS: *Pisum sativum* L.

Origin and types

The modern cultivars have been developed from material originally introduced into Africa, China, Europe and India from South-West Asia. This early distribution over a wide area is generally believed to be the reason for the present diversity of types. Peas are now widely cultivated in temperate regions and as a cool season crop in the tropics, especially at the higher altitudes.

Table 11.1 The important species of tropical vegetable legumes with common names, sowing rates, mean seed yields and 1000 grain weights (after Grubben, 1977)

Species	Common names	Sowing rate (kg/ha)	Mean seed yield (tonnes/ha)	1000 grain weight (g)
Arachis hypogea L.	Groundnut	80	1.0	625
Cajanus cajan (L.) Millsp.	Pigeon pea	15	1.2	125
Canavalia gladiata (Jacq.) D.C.	Sword bean	60	1.3	4000
Lablab niger Medik.	Hyacinth bean, Lablab bean	40	0.9	330
Glycine max (L.) Merr.	Soya bean	50	1.3	670
Pachyrrhizus erosus (L.) Urban	Yam bean	30	0.6	250
Phaseolus aureus Roxb. syn. *Vigna radiata* L. Wilczek var *radiata*	Green or Golden gram, Mung bean	5	0.5	30
Phaseolus lunatus L.	Lima bean	50	1.0	300
Psophocarpus tetragonolobus (L.) D.C.	Winged bean	20	1.1	500
Vigna unguiculata (L.) Walp.	Cowpea, Yardlong bean	20	0.7	220
Voandzeia subterranea (L.) Thou.	Bambara groundnut	40	0.6	670

Some authorities have recognized two species, *Pisum arvense* L. (field peas) and *P. sativum* L. (garden peas). A further subdivision of *P. sativum* L. has also been used, i.e. *P. sativum* var. *macrocarpon* Ser. (edible podded sugar peas) and *P. sativum* var. *humile* Poir (early dwarf peas in which the pods are lined with a characteristic membrane). However, all the types are now generally considered to be within the species *P. sativum* which is divided into two main groups, field peas and vegetable peas. The field peas include pigeon and maple peas with many cultivars specifically developed for fodder (including silage), forage and animal feed according to criteria such as vegetative habit and protein content of their seeds. Some of the field pea cultivars are grown for the production of grain (i.e. dried peas) used for the production of processed peas and split peas in addition to traditional dried peas.

The processing of peas is a highly developed industry in some temperate areas where the immature peas are either canned or frozen, or the mature dried peas are subsequently processed and canned. Each of these outlets requires cultivars with specific characters such as suitability for early sowing, time of maturity, seed size and colour.

Cultivars are also available with specific characters which suit them to commercial or private garden production of the immature seeds. Important characters include suitability for mechanical harvesting ('vining'), early production, yield, flavour and colour of the immature seeds.

Cultivar description of pea

Season of use and suitability for specific outlets (e.g. processing)

Seed:	round or wrinkled
	relative size
	colour (influenced by colour of testa and cotyledons which can be recorded separately)
Plant height:	number of nodes to first flower
Stem:	presence or absence of pigmentation
	degree of fasciation
Leaf:	leaflets – relative size, colour, marbling
Stipules:	developed or vestigial
	extent of marbling
Tendrils:	degree of development
Flower:	number per raceme
	colour
Pod:	relative colour
	relative length
	relative shape
	form at tip (apex)
	degree of parchment

Resistance to specific pathogens, e.g. *Fusarium oxysporum* and *Erysiphe polygoni*

Soils and nutrition

Peas are grown on soils with a pH 5.5–6.5. The ratio of fertilizers applied in the base dressing during seed bed preparations depends on the nutrient status and local customs but specialist pea seed producers tend to use a N:P:K ratio of 3:1:2 which is a higher nitrogen level than that normally used for the market crop. Recent work by Browning and George (1981a) has indicated that increased seed yield can be obtained with relatively high levels of N and P. In addition, seeds produced from the higher N and P mother plant regimes were found on analysis to contain higher levels of N and P. However, the seeds from the higher nitrogen regimes in the same experiments were found to be less vigorous when subjected to the conductivity test (PGRO 1978). This indicates that nutrient regimes for seed yield differ from those required to produce high quality seed.

Peas are very susceptible to manganese deficiency especially on wet soils with relatively high organic levels. The symptoms of the deficiency are brownish hollow centres to the cotyledons seen when unripe seeds are split open. The condition is commonly called 'marsh spot'. Manganese sulphate is included in base dressings at a rate of 40–100 kg/ha where manganese deficiency is known to occur. Alternatively, manganese sulphate is applied as a foliar spray at a rate of 10 kg/ha in 200–1000 litres of water but this must be applied as soon as the symptoms are diagnosed otherwise it will be too late to improve seed yield and quality.

Another indication of low vigour in pea seeds is the incidence of 'hollow heart' and 'bleaching'. The former is characterized by a cavity of dead tissue on the adaxial surface of the cotyledons (Fig. 11.1) and later by a yellowing of the green seeds as a result of chlorophyll bleaching (Maguire *et al* 1973).

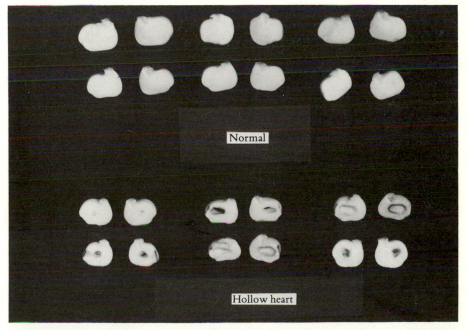

Fig. 11.1   Pea seeds with hollow heart compared with normal seeds.

It is generally accepted that both these conditions are caused by high temperatures during maturity or seed drying (Perry and Harrison 1973). During the course of the investigations on mother plant nutrition by Browning and George (1981b) it was shown that the low nitrogen and high phosphorus regime predisposed developing seeds to hollow heart whereas the high nitrogen regimes predisposed them to bleaching.

Irrigation

Salter and Goode (1967) reviewed the literature relating to the irrigation of peas and found that most workers have reported a moisture-sensitive stage during flowering which lasts until petal fall. This moisture-sensitive stage is very evident regardless of the amount of available water before or after flowering. Moisture deficits prior to anthesis only affect the weight of haulm produced but shortage of water during pod growth can also affect yield.

The general practice in temperate regions is, therefore, to ensure that adequate irrigation is available during and after flowering, whereas in arid areas and dry seasons in the tropics sufficient water should also be given to ensure satisfactory plant growth before flowering commences.

Sowing

Pea seeds are direct drilled into the field relatively early in the spring at the rate of 150 kg/ha in rows 40 cm apart although sowing rate and plant density used depend on the seed size and growth habit of the cultivar. Some North American seed producers use a sowing density of up to 250 kg/ha with a much shorter distance between rows for the final multiplication stage. The higher plant densities reduce the time span of pod maturity but increase the difficulty of roguing.

Flowering

This species is a quantitative long-day plant. The earliest cultivars generally flower on lower nodes than the taller and later types.

Pollination and isolation

Pea flowers are almost totally self-pollinated. This occurs in the late bud stage before the flower is completely open.

Recommended isolation distances for peas are relatively short in many countries and aim mainly to prevent admixture during harvesting. It is important that adjacent cultivars should be at least 20 m apart with the distance increased to at least 100 m for stock seed production.

Roguing stages

1. *After emergence* (when plants are approximately 15 cm high), remove plants which are taller off-types. For basic seed production particular attention is given to checking that foliage, including stipules, is typical of the cultivar (Fig. 11.2).

2. *Flowering* Remove early flowering plants from late flowering cultivars, check flower colour and remove any plants with flower colour which is not true to type. Check flower number per node.

3. *Pods formed* Check that pod shape, size and colour are typical of the cultivar (Fig. 11.2); remove late flowering plants; remove non or low-yielding plants.

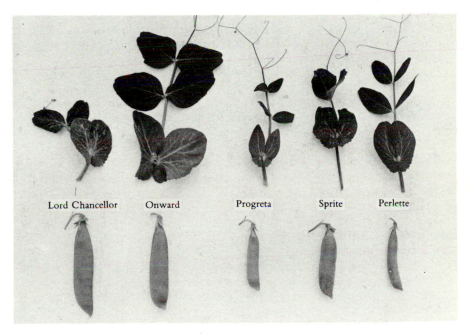

Lord Chancellor   Onward   Progreta   Sprite   Perlette

Fig. 11.2   Pea cultivars showing range of stipule, leaf, tendril and pod characters: left to right 'Lord Chancellor', 'Onward', 'Progreta', 'Sprite' and 'Perlette'.

Harvesting

The harvesting operation normally commences when the majority of pods have become parchment-like. By this stage the maturity of individual seeds is sufficiently advanced for them not to be affected by subsequent drying. Biddle and King (1977) showed that the seed quality is reduced if seeds are harvested when their moisture content is above 30–35 per cent.

Combining

In areas where the seeds dry sufficiently on the plants the crop is harvested by direct combining . A relatively low drum speed is used and care taken to avoid mechanical damage to the seeds.

Desiccants, such as Diquat, are used by some pea seed producers in areas where the pre-harvest drying-off of the plant material is relatively slow. They are applied when the crop starts to senesce and the lowest pods have turned pale brown and parchment-like. The usual rate of application is 3 litres in 100 litres of water sprayed on at the stage when random samples of seeds have a moisture content of 40 per cent.

Crops which have been treated with a desiccant are either combined direct when the moisture content of the seeds is approximately 28 per cent or cut into windrows four or five days following application of the desiccant.

If combining is not possible due to unavailability of machines or the production scale is too small, an alternative is to cut the crop when the earliest peas have dried to the parchment stage and the foliage is starting to dry off, characterized by a reduction in the intensity of green in the leaves and haulm. Sample pods should contain fully developed seeds which are firm, taste starchy and are readily detachable from their pods.

Windrowed crops are usually turned, although each time the windrows are turned some seed is lost by shattering.

An alternative to windrows is to cut and stack the crop on to tripods. This increases the rate of drying and reduces the time in which seeds can deteriorate as a result of pathogen activity but the method has a relatively high labour content.

Material from windrows or tripods is threshed when the seeds' moisture content is approximately 30 per cent using a thresher with rubber covered tines.

Seed yield and 1000 grain weight

The average yield of pea seed is approximately 2000 kg/ha.

The 1000 grain weight varies from the small to the larger seeded types. But a guideline is 330 g for the larger seeded types to 150 g for the smaller ones.

The main seed-borne pathogens of peas with common names of the diseases they cause

Pathogens	Common names
Ascochyta pisi Lib.	Leaf and pod spot
Botrytis cinerea Pers. ex Pers.	Grey mould, pod rot
Cladosporium cladosporiodes (Fresen.) de Vries f.sp. *pisicola* (Snyder) de Vries	White mould, leaf and stem spot
Colletotrichum pisi Pat.	Anthracnose
Erysiphe pisi DC. ex St. Am.	Powdery mildew
Fusarium oxysporum Schlecht. ex Fr. f.sp. *pisi* (van Hall) Snyd. and Hans	Wilt

Pathogens	Common names
Fusarium spp.	Wilt
Mycosphaerella pinodes (Berk. and Blox.) Vesterg., syn. *Ascochyta pinodes* Jones	Foot rot, leaf spot, black spot
Perenospora viciae (Berk.) Casp., syn. *P. pisi* (de Bary) Syd.	Downy mildew
Phoma medicaginis Malbr. and Roum. var. *pinodella* Jones, syn. *Ascochyta pinodella* Jones	Foot rot, collar rot
Pleospora herbarum (Pers. ex Fr.) Rabenh., syn. *Stemphylium botryosum* Wallr.	Foot rot
Rhizoctonia solani Kühn	Damping-off, stem rot
Sclerotinia sclerotiorum (Lib.) de Bary	Stem rot
Septoria pisi West	Leaf blotch, *Septoria* blotch
Pseudomonas phaseolicola (Burkholder) Dowson, syn. *Pseudomonas medicaginis* Sackett var. *phaseolicola* (Stapp and Kotte) Dowson	Bacterial blight
Pseudomonas pisi Sackett	Bacterial blight
Xanthomonas rubefacines (Burr) Magrou and Prévot	Purple spot
Viruses	Pea early browning virus, Pea enation virus, syn. *Pisum* virus 1 Pea mild mosaic Pea mosaic virus Pea seed-borne mosaic virus (syns. pea fizzle-top virus and pea leaf rolling mosaic virus)

FRENCH BEAN: *Phaseolus vulgaris* L.

Phaseolus vulgaris, which originated in America, is cultivated throughout the temperate, tropical and sub-tropical areas of the world. The evolution of *Phaseolus* species under domestication is discussed by Smartt (1976).

There are several common names for this crop and, in addition to French bean, it is also referred to as common bean, snap bean, green bean, kidney bean, haricot bean and dwarf bean. The majority of cultivars are bush types but there are also climbing types which are usually referred to as 'climbing French bean'. This species is cultivated for either dried beans or the immature green pods. The species is an important crop for processing, the immature green pods are canned or frozen; the dried beans are cooked in tomato purée and canned, a product frequently known as navy or baked beans. Specific cultivars have been developed for each of these processes and 'stringless' types have also been selected which are suitable for producing the fresh crop and for processing as green beans.

Cultivar description of French bean

Season and use   flowering and cropping season, suitability for specific market outlets, e.g. green, processing or dried, suitability for mechanical harvesting

Seed   relative length
                  colour of testa; colours of testa and spots, bars or splashes if bi-coloured
                  shape
                  resistance to mechanical damage

Plant habit   dwarf or climbing, general habit of bush types with degree of branching

Leaf   shape, texture, colour and relative size of mature leaf

Flower   colour

Pod   relative length and shape, degree of curvature colour, external pigmentation
                  suture – presence or absence of string
                  wall – presence or absence of parchment transverse section, shape

Resistance to specific pathogens, e.g. *Colletotrichum lindemuthianum* (Sacc. and Magn.) Bri. and Cav. anthracnose) and bean yellow mosaic virus

Nutrition

Phaseolus vulgaris tolerates slightly acid soil conditions and soils with a pH between 5.5 and 6.5 can be used successfully. The general N : P : K ratio applied during seed bed preparation is 1 : 2 : 2. Little work has been done until recently on the effect of mother plant mineral nutrition on seed yield and quality. Work by Gavras (1981) has indicated different ratios of nitrogen and phosphorus were required for relatively high seed yield and for seed quality. Browning *et al.* (1982) reported the importance of a correct balance between nitrogen and phosphorus available to the plant for the production of high vigour seeds. Figure 11.3 shows the effects of three levels each of phosphorus and nitrogen on the yield of bean seeds and the quality of seeds as determined by the cold test. The cold test is described by Perry (1981).

Irrigation

The majority of irrigation experiments reviewed by Salter and Goode (1967)

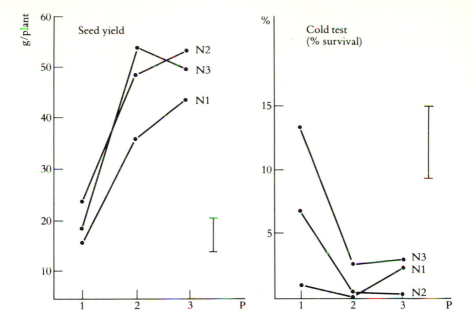

Fig. 11.3   Effects of three phosphorus and three nitrogen levels on seed yield and
seed vigour in *Phaseolus vulgaris* L.

indicated that water shortages during flowering and pod development
seriously affected the yield of beans. All the experimental evidence indicated
that irrigating at these two stages gave the maximum benefit. Work in European temperate regions showed that irrigation applied before flowering
commences only increases vegetative growth. However, the findings of
North American workers suggest that under drier soil conditions water applied during the vegetative stage is also beneficial to yield.

Sowing

Seed of the bush types is sown in the late spring when the possibility of frost
has passed and soil temperatures have risen sufficiently for satisfactory germination and plant growth. In some specialized areas of North America,
notably Michigan where the length of season allows double cropping, seeds
of early-maturing cultivars are drilled into the stubble of a previous barley
crop or immediately after ploughing in the stubble. When this technique is
adopted seeds of early-maturing cultivars are drilled in early July.

The sowing rate depends on the relative seed size of the cultivar but is
approximately 100 kg/ha for the bush types and 50 kg/ha for the pole types.
Bush cultivars are sown in rows 45–90 cm apart, according to cultivar and
stage of multiplication. The climbing cultivars are sown in rows 90–120 cm
apart. In some tropical areas cane frames or supports are erected, although
generally they are supported by strings.

Flowering and pollination

Most cultivars of each type (i.e. bush and climbing) are day-neutral although there are also some short-day cultivars in each group.

The flowers are self-compatible and are predominantly self-pollinated although some cross-pollination occurs. The degree of cross-pollination probably increases in the tropics where the activity of insects (including thrips) is greater (Drijfhout 1981).

Isolation

Most authorities stipulate a distance of 50 m between *Phaseolus vulgaris* cultivars for the final stage of multiplication but a minimum of 150 m is recommended for basic seed production.

Roguing stages

1. *Before flowering*   Check plant habit, vigour and height according to type; check foliage, leaf shape and colour.

2. *At flowering*   Check plant vigour and flower colour; remove plants showing symptoms of seed-borne pathogens.

3. *Seed set and pods formed*   Check pod characters, including shape, colour; remove late flowering off-types and plants showing symptoms of seed-borne pathogens.

Basic seed production

A scheme for maintaining breeders' stocks of beans has been described by Drijfhout (1981). For basic seed single plant selections are made from a relatively large plant population grown from breeders' stock seed. The selections are based on five successive inspections commencing immediately after seedling emergence. Particular care is taken to retain only seedlings with appropriate cotyledon colour. The young plants are then inspected a second time to remove plants which show inappropriate leaf characters. The third selection is made just prior to flowering in order to confirm the general plant form, height and earliness.

When flowering commences the plants are examined a fourth time to check flower colour, off-types are again removed. The fifth selection is made when the pods are developing and characters such as pod shape, length and colour can be seen, and other characters such as stringlessness can be detected. Plants showing symptoms of seed-borne pathogens are discarded at each stage of selection.

The remaining plants are harvested singly and low yielding plants (as determined by seed yield) are rejected. The remainder of seeds from selected plants are sown the following season as single plant selections and undesirable lines are rejected. The remainder are bulked for further multiplication and used as basic stock.

Harvesting

The dwarf or bush types are generally considered to be ready for a once over harvest when the earliest pods are dry and parchment-like and the remainder of the pods have turned yellow. Seed maturity is confirmed by opening sample pods; the seeds should be fully developed with a mealy texture. Under good growing conditions the flowers tend to set until relatively late in the season. This leads to a loss from 'shattering' of the earliest mature seeds if harvesting is delayed. Smith (1955) examined the effects of stage at harvesting on bean seed yield and quality and found that there was a reduction of 318 lb/acre (358 kg/ha) in harvested seed when the crop was cut before the earliest pods were fully mature.

The plants are either cut and placed in windrows for further drying before either combining or threshing, or they are combined direct from the standing crop. The entire operation is planned to ensure both the minimum loss from shattering and the least possible mechanical damage to the seeds which are especially susceptible to cotyledon cracking. Figure 11.4 shows plants for basic seed production dried *in situ* in Kenya where satisfactory weather conditions allow the maximum number of pods to mature and the minimum loss of seeds from shattering.

The climbing cultivars mature over a longer period than the bush types so they are harvested by hand on three or more successive occasions as the older pods mature or the plants are pulled out and dried off in windrows.

Threshing

Small quantities of seed are threshed by hand to avoid subsequent loss due to mechanical damage; this is especially important with small lots of basic or stock seed. Large lots are threshed with a drum speed of 250–350 rpm at a concave clearing of *c*. 12–20 cm. The seeds' moisture content should not be too low or excessive mechanical damage occurs during machine threshing.

Seed yield and 1000 grain weight

The seed yield is *c*. 1500 kg/ha although under ideal production and harvesting conditions yields are *c*. 2000 kg/ha.

The 1000 grain weight for *Phaseolus vulgaris* is *c*. 250 g, although this can be up to *c*. 600 g in the smaller seeded cultivars.

Fig. 11.4   Dwarf bean plants with ripe pods prior to harvesting.

The main seed-borne pathogens of *Phaseolus* spp. with common names of the diseases they cause

Pathogens	Common names
Ascochyta spp.	*Ascochyta* leaf spot
Aspergillus sp.	Baldhead and snakehead of seedlings
Botrytis cinerea Pers. ex Pers.	Grey mould, pod rot
Cercospora sp.	Leaf blotch
Colletotrichum lindemuthianum (Sacc. and Magn.) Bri. and Cav.	Anthracnose
C. phaseolorum Tak.	Anthracnose of Adzuke bean
C. truncatum (Schw.) Andrus and Moore	Stem anthracnose of lima bean

Pathogens	Common names
Diaporthe phaseolorum (Cook and Ell.) Sacc. var. *Sojae* (Lehm.) Wehm.	Pod and stem blight of lima bean
Elsinoë phaseoli Jenkins	Scab of lima bean
Fusarium oxysporum Schlecht. ex Fr. f.sp. *phaseoli* Kendr. and Snyd.	Yellows and wilt of *P. vulgaris*
F. solani (Mart.) Sacc. f.sp. *phaseoli* (Burkh.) Snyd. and Hans.	Root rot
Macrophomina phaseolina (Tassi.) Goid., syns. *M. phaseoli* (Maubl.) Ashby *Rhizoctonia bataticola* (Taub). Briton-Jones *Sclerotium bataticola* Taub.	Ashy stem blight and charcoal rot
Phaeoisariopsis griseola (Sacc.) Ferraris, syn. *Isariopsis griseola* Sacc.	Angular leaf spot of *P. vulgaris*
Phytophthora phaseoli Thaxter	Downy mildew and collar rot of lima bean
Pleospora herbarum (Pers. ex Fr.) Rabenh. syn. *Stemphylium botryosum* Wallr.	Red nose, leaf spot
Rhizoctonia solani Kühn	Damping-off, stem canker
Sclerotinia sclerotiorum (Lib.) de Bary	*Sclerotinia* wilt, stem rot, watery soft rot, white mould
Uromyces appendiculatus (Pers.) Unger	Rust
Corynebacterium flaccumfaciens (Hedges) Dowson	Bacterial wilt
Corynebacterium sp.	Brown stem
Pseudomonas phaseolicola (Burkholder) Dowson, syn. *P. medicaginis* Sackett var. *phaseolicola* (Burkholder) Stapp and Kotte	Halo blight, grease spot
Pseudomonas syringae van Hall	Bacterial brown spot of *P. lunatus* and *P. vulgaris*
Xanthomonas phaseoli (E. F. Smith) Dowson	Common bacterial blight of *P. vulgaris*
X. phaseoli (E. F. Smith) Dowson var. *fuscans* (Burkholder) Starr and Burkholder	Fuscous blight of *P. vulgaris*
Viruses	Bean common mosaic virus (syn. bean mosaic virus) of *P. lunatus* and *P. vulgaris*
	Black grain leaf crinkle virus
	Broad bean mottle virus of *P. mungo*
	Cherry leaf roll virus of *P. vulgaris*
	Cucumber mosaic virus of *P. aureus*
	Runner bean mosaic virus

RUNNER BEAN: *Phaseolus coccineus* L. (syn. *P. multiflorus* Willd.)

Origin and types

The runner bean, also commonly called 'scarlet runner bean' originates from Central America but is now widely distributed in the temperate regions. It is especially popular in Europe where it is mainly cultivated as a garden vegetable. Although a perennial with a tuberous rootstock, because of its sensitivity to frost it is cultivated as an annual.

The main product is its edible partly developed green pods, but in some areas the mature seeds are also used for culinary purposes.

Most cultivars are of the twining form which is normally grown on supports. There are a few non-climbing types, a notable cultivar is 'Hammonds Dwarf Scarlet'. The dwarf forms are mainly used for field crop production.

Cultivar description of runner bean

Use	fresh market, processing, private garden (including suitability for exhibition)
Plant habit	climbing or dwarf
Flower colour	red, white or bicolour
Pod characters	relative length, texture, colour 'string' or 'stringless' types longevity of pods in marketable condition while attached to the plant
Seed colour	black, purple, pink, white or bi-colour (e.g. pink with purple specks)

Agronomy

The soil and cultivar requirements are the same as described above for *Phaseolus vulgaris*. There is evidence (Blackwall 1961, 1962, 1963) that runner beans respond to organic manures. Seed is sown at a rate of up to 100 kg/ha when grown on poles or strings in rows which are 100 cm apart. The method of training for seed production varies from one area to another. In large-scale production of the climbing types the final multiplication is done without plant supports, but for basic seed production, plants are grown on individual strings to facilitate roguing. The dwarf types are grown at closer row spacings similar to *P. vulgaris* but with approximately 50 per cent increase in sowing rate because of the larger seeds of *P. coccineus*.

Irrigation

The main work on the response of *Phaseolus coccineus* to different soil moisture conditions at different stages of growth has been done by Blackwall (1961, 1962, 1963). Some of Blackwall's investigations included a study of the interaction of moisture status with organic manure treatments. The application of farmyard manure during soil preparations followed by irrigation from the visible early green flower bud stage increased the early yield of green beans by 57 per cent. Withholding irrigation until the flowers showed their colour increased yields by only 16 per cent. These increases were obtained in comparisons with runner beans grown in plots with no additional farmyard manure or irrigation.

Flowering

The runner bean is a long-day plant but does not set pods when temperatures are more than 25 °C. It is also similarly sensitive to night temperatures below 20 °C which tend to cause abscission of both flowers and young pods (Smartt 1976).

Pollination

Phaseolus coccineus is self-fertile but some insect activity is required to 'trip' the stigma (i.e. rupture the stigmatic surface slightly). This is usually achieved by honeybees (*Apis mellifera*) or bumble bees (*Bombus* species). There is up to 40 per cent cross-pollination, especially early in the flowering season. Some species of bumble bees (*Bombus lucorum* and *B. terrestris*) are short-tongued and obtain nectar by making holes near the base of the flowers (Free and Racey 1968). They are not therefore efficient cross-pollination agents.

Isolation

A minimum isolation distance of 100 m is stipulated by most authorities, although this distance should be increased for stock seed production. Some authorities require a greater isolation distance between cultivars with different flower colours.

Roguing

The roguing stages and characters to observe during this operation are similar to those described above for *P. vulgaris*. Particular attention should be given to the early removal of off-types before they have contributed to the pollination of other plants.

Harvesting and threshing

The harvesting methods are similar to those described for *P. vulgaris* although a greater degree of hand harvesting it done from plants grown on supports.

Runner beans are threshed with a drum speed of 250–350 rpm and a wide concave opening to minimize mechanical damage.

Seed yield and 1000 grain weight

The average seed yield is *c*. 1000 kg/ha. The 1000 grain weight is *c*. 1 kg, although the weight of some smaller seeded cultivars is less.

Seed-borne pathogens

The main seed-borne pathogens of *Phaseolus* species are listed at the end of the section on *P. vulgaris*.

BROAD BEAN: *Vicia faba* L.

Origin and types

This species is also commonly referred to as field bean, horse bean, tick bean and Windsor bean although all but the latter common name usually refer to cultivars not used as vegetables for human consumption (see below).

Vicia faba originates from the Mediterranean region and South-West Asia. It is widely grown in temperate areas and in the tropics, sub-tropics and arid areas as a cool season crop.

Vicia faba L. was originally divided into three subspecies. *Vicia faba* L. var. *minor* Beck and *Vicia faba* L. var. *equina* Pers. were considered the tick bean and horse bean respectively despite overlap of types, and the broad bean was referred to as *Vicia faba* L. var. *major* Harz. In earlier classifications seed size was one of the main criteria for division into basic types but this is now no longer accepted as there are some small seeded broad beans and large seeded horse beans. The main classification of cultivars is now based on their use. Thus the broad beans used as a vegetable are considered separately from the field beans grown for fodder and animal feed which are subdivided into tick beans and horse beans. Broad beans are used as a fresh vegetable in temperate areas where there is also an increase in the use of the fresh immature seeds for canning and freezing. In some arid and tropical areas the dried mature seeds are stored for use during the winter or other seasons when fresh vegetables are in short supply.

Cultivar description of broad bean

Season and use	suitability for early sowing, winter hardiness and maturity period
Seed	weight of 1000 seeds relative size colour of testa shape hilum, colour
Plant	height number of pod bearing tillers
Leaf	stipules, presence or absence of pigmented area
Flower	wing petals, presence or absence of pigmented spot on standard petal, presence or absence of spot on back
Pod	length breadth approximate number of seeds per pod

Soil and nutrition

Soils for broad beans should have a pH of 6.5 with adequate available calcium. A suitable liming material must be applied during soil preparations.

The nutrient ratio applied during the final stages of seed bed preparations is N : P : K 1 : 1 : 1. It is particularly important not to apply excessive nitrogen, especially for autumn sown crops.

Sowing

The longpod cultivars which have some frost resistance are sown in the late autumn but in areas with severe winters sowings are made in early spring as with the other types of broad bean.

Seeds are sown at the rate of 150 kg/ha in double rows 25 cm apart with 70 cm between the double rows.

Irrigation

Broad beans are very responsive to irrigation applied during anthesis. Brouwer (1949, 1959) found that dry soil at this stage had the greatest adverse effect on yield while irrigation before the onset of flowering had relatively little advantage even under dry conditions. However, Jones (1963) found that seed yield was to some extent dependent on the plants having a relatively high growth rate before flower development. It is therefore better to ensure that the plants receive sufficient water both during early development and from the start of flowering.

Flowering

The majority of broad bean cultivars require long days for the acceleration of flower initiation although the earliest flowering cultivars do not (Whyte 1960). There is also evidence that some vernalization at temperatures below 14 °C will also accelerate flowering.

Pollination

Broad beans are self-compatible but both self- and cross-pollination occur. Insect activity is responsible for up to 30 per cent crossing (Watts 1980). However, as with *Phaseolus coccineus* some bees obtain nectar by making a hole at the base of the flowers as the season advances, thus reducing the incidence of cross-pollination.

Reisch (1952) found that high rainfall during flowering prevented pollination by insects and thereby was responsible for a considerable reduction in yield.

Isolation

Most authorities stipulate a minimum isolation distance of 300 m between broad bean crops. However, because of the relatively high incidence of cross-pollination, isolation distances should be at least 1000 m, especially for stock seed production.

Roguing stages

1. *Before flowering commences*
   (a) General plant habit, vigour, height, tiller number.
   (b) Stipules: presence or absence of pigment spot.
   (c) Remove plants showing symptoms of seed-borne pathogens, e.g. *Ascochyta fabae*.

2. *At early flowering*
   (a) General plant habit
   (b) Flower colour: individual cultivars have some specific markings on parts of flower, e.g. on wing petals or standard.
   (c) Remove plants showing symptoms of seed-borne pathogens

3. *When pods set*
   (a) Colour, shape and relative length of pods.
   (b) Pose of pods.

Harvesting and processing

Broad beans are ready for harvest when the pods have become relatively dry and lost their sponginess; this stage is usually preceded by a general blackening of the pods.

The crop is either cut by hand and tied in bundles or mown by machine. Bundling and stooking is necessary in Northern Europe although not in the drier production areas such as the Middle East and North Africa.

The material is threshed when dry at a drum speed of 250 rpm. Care must be taken to adjust the concave setting according to the seed size of the cultivar to minimize mechanical damage to the seeds.

Seed yield and 1000 grain weight

The seed yield is $c.1500$ to 2000 kg/ha. The 1000 grain weight is $c.800–1200$ g for the smaller seeded types and is $c.2000$ g for the larger seeded types.

The main seed-borne broad bean pathogens with common names of the diseases they cause

Pathogens	Common names
Ascochyta fabae Speg.	Leaf and pod spot
Botrytis fabae Sardiña	Chocolate spot
Colletotrichum lindemuthianum (Sacc. and Magn.) Briosi and Cav.	Anthracnose
Fusarium spp.	*Fusarium*
Pleospora herbarum (Pers. ex. Fr.) Rabenh., syn. *Stemphylium botryosum* Wallr.	Net blotch
Uromyces viciae-fabae (Pers.) Schroet., syn. *Uromyces fabae* (Grev.) de Bary	Rust
Pseudomonas fabae (Yu) Burkholder	
Viruses	Bean yellow mosaic virus
	Broad bean mild mosaic virus (syn. bean seed pattern virus)
	Broad bean stain virus
	Echtes ackerbohnenmosaik virus (syn. broad bean true mosaic virus)
	Broad bean wilt virus
	Pea seed-borne mosaic virus (syn. pea fizzle-top virus)
Ditylenchus dipsaci (Kühn) Filipjev	Stem eelworm

REFERENCES

Biddle, A. J. and King, J. M. (1977) Effect of harvesting on pea seed quality, *Acta Horticulturae* **83**, 77–81.

Blackwall, F. L. C. (1961) Factors affecting the set of runner bean pods, *A. R. Nat. Veg. Res. Sta., Wellesbourne, 1960*, 40.

Blackwall, F. L. C. (1962) Factors affecting the set of runner bean pods, *A. R. Nat. Veg. Res. Sta., Wellesbourne, 1961*, 48–9.

Blackwall, F. L. C. (1963) Factors affecting the set of runner bean pods, *A. R. Nat. Veg. Res. Sta., Wellesbourne, 1962*, 49.

Brouwer, W. (1949) Steigerung der Erträge der Hülsenfrüchte durch Beregnung sowie Fragen der Bodenuntersuchung und Düngung, *Z. Acker-u. PflBau* **91**, 319–46.

Brouwer, W. (1959) *Die Feldberegnung.* DLG-Verlag, Frankfurt/Main, 4th edn.

Browning, T., Gavras, M. and George, R. A. T. (1982) *Proceedings XXI International Horticultural Congress, Hamburg.*

Browning, T. H. and George, R. A. T. (1981a) The effects of nitrogen and phosphorus on seed yield and composition in peas, *Plant and Soil* **61**, 485–8.

Browning, T. H. and George, R. A. T. (1981b) The effects of mother plant nitrogen and phosphorus nutrition on hollow heart and bleaching of pea (*Pisum sativum* L.) seed, *J. Exp. Bot.* **32**(130), 1085–90.

Drijfhout, E. (1981) Maintenance breeding of beans. *Proceedings of FAO/SIDA Technical conference on improved seed production, Nairobi, Kenya.*

Free, J. B. and Racey, P. A. (1968) The pollination of runner beans (*Phaseolus multiflorus*) in a glasshouse, *J. Apicult. Res.* **7**(2), 67–9.

Gabelman W. H. and Williams, D. D. F. (1960) Developmental studies with irrigated snap beans, *Res. Bull. Wis. Agric. Exp. Sta.* **221**.

Gavras, M. (1981) The influence of mineral nutrition, stage of harvest and flower position on seed yield and quality of *Phaseolus vulgaris* L. Ph.D. Thesis, University of Bath, England.

Grubben, G. J. H. (1977) *Tropical Vegetables and Their Genetic Resources*, H. D. Tindall and J. T. Williams (eds). International Board for Plant Genetic resources, IBPGR/FAO, Rome.

Jones, L. H. (1963) The effect of soil moisture gradients on the growth and development of broad beans (*Vicia faba* L.), *Hort. Res.* **3**, 13–26.

Maguire, J. D., Kropf, J. P. and Steen, K. M. (1973) Pea seed viability in relation to bleaching, *Proc. Ass. Off. Seed Anal.* **63**, 51–58.

Perry, D. A. (Ed.) (1981) *Handbook of Vigour Testing Methods.* International Seed Testing Association, Zurich.

Perry, D. A. and Harrison, J. G. (1973) Causes and development of hollow heart in pea seed. *Ann. Appl. Bid.* **73**(1), 95–101.

PGRO (1978) Information sheet No. 70. Processors' and Growers' Research Organisation.

Reisch, W. (1952) Variabilitätsstudien an *Vicia faba* L., *Z. Acker-u PflBau* **94**, 281–306.

Salter, P. J. and Goode, J. E. (1967) *Crop Responses to Water at Different Stages of Growth.* Commonwealth Agricultural Bureaux, Farnham Royal, England.

Smartt, J. (1976) *Tropical Pulses.* Longman, London.

Smith, F. L. (1955) The effects of dates of harvest operations on yield and quality of pink beans, *Hilgardia* **24**, 37–52.

Watts, L. E. (1980) *Flower and Vegetable Plant Breeding.* Grower Books, London.

Whyte, R. O. (1960) *Crop Production and Environment.* Faber and Faber, London.

FURTHER READING

Bianco, V. V., Miccolis, V. and Damato, G. (1981) Influence of plant density on seed yield of three cultivars of *Vicia faba* L., *Acta Horticulturae* **111**, 209–16.

Gaunt, R. E. and Liew, R. S. S. (1981) Control of diseases in New Zealand broad bean seed production crop, *Acta Horticulturae* **111**, 109–14.

Gavras, M. and George, R. A. T. (1981) A review of the effects of mineral nutrition on the seed yield and quality in *Phaseolus vulgaris*, *Acta Horticulturae* **111**, 191–4.

Higgins, J., Evans, J. L. and Reed, P. J. (1981) Classification of Western European cultivars of *Vicia faba* L. *J. Nat Inst. Agric. Bot.* **15**, 480–7

Kooistra, E. (1964) Identification research on pulses, *Proc. Int. Seed Test. Ass.* **29**(4), 937–46.

Linsalata, D. and Bianco, V. V. (1981) Effects of time of sowing and water regime on seed yield of broad beans, *Acta Horticulturae* **111**, 203–8.

Matthews, S. (1973) The effect of time of harvesting on the viability and pre-emergence mortality in soil of pea (*Pisum sativum*) seeds, *Ann. Appl. Biol.* **73**, 1–9.

Peat, W. E. (1982) Reproductive losses in the faba bean, *Outlook on Agriculture* **11**(4), 179–84.

Powell, A. A. and Matthews, S. (1980) The significance of damage during imbibition to the field emergence of pea (*Pisum sativum* L.) seeds, *J. Agric. Sci., Camb.* **95**, 35–38.

Powell, A. A. and Matthews, S. (1981) The significance of seed coat damage in the production of high quality legume seeds, *Acta Horticulturae* **111**, 227–34.

Salter, P. J. and Drew, D. H. (1965) Root growth as a factor in the response of *Pisum sativum* L. to irrigation, *Nature* **206**, 1063–4.

Sneddon, J. L. (1969) Identification features of white seeded stringless varieties of French beans (*Phaseolus vulgaris* L.), *J. Nat. Inst. Agric. Bot.* **11**, 476–98.

Sutcliffe, J. F. and Pate, J. S. (Eds.) (1977) *The Physiology of the Garden Pea.* Academic Press, London and New York.

Tucker, C. L. and Harding, J. (1975) Outcrossing in common bean *Phaseolus vulgaris* L., *J. Amer. Soc. Hort. Sci.* **100**(3), 283–5.

12 SOLANACEAE

The genera in this family provide some of the most important vegetables in the world, viz:

Lycopersicon esculentum Mill.	tomato
Solanum tuberosum L.	potato
Solanum melongena L.	aubergine, eggplant
Capsicum annum L.	sweet pepper
Capsicum frutescens L.	chilli pepper

There are several other vegetable species in this family including *Physalis peruviana* L., cape gooseberry, and *Cyphomandra betacea* (Cav.) Sendt., tree tomato, but these are of local importance only and their commercial seed production is therefore not discussed, although it is noted that commercial production of fruits of *Physalis* spp. is increasing in several parts of the world and they may subsequently become important as a seed crop.

There are a few *Solanum* species collectively known as 'African eggplants' which are mainly used as leaf vegetables, especially locally in parts of Africa. These include *Solanum macrocarpon* L. which is important in West Africa, and *S. aethiopicum* L. used as a leafy vegetable in Central Africa. Grubben (1977) discusses the role of these and better known *Solanum* spp. and indicates the need for collecting and screening local cultivars. As far as is known to the author there is no commercial seed production of these two *Solanum* spp.

The five main solanaceous vegetables listed above will be discussed individually. There are in addition some other important crop plants in this family which includes *Nicotiana tabacum* L. (tobacco) in addition to some drug plants, such as *Atropa belladonna* L. and *Datura stramonium* L. Although these last three species are clearly not vegetables, it is as well to realize that as related crops they are important alternative hosts to the viruses which infect a wide range of genera in Solanaceae and should be taken into account when considering isolation requirements.

THE TOMATO: *Lycopersicon esculentum* Mill.

Origins and types

The species of the genus *Lycopersicon* originate in the narrow west coast area of South America extending from Ecuador to Chile, between the Andes and the sea, except *L. cheesmanii* Riley, which is found in the Galapagos Islands. Evidence indicates that material of *L. esculentum* Mill. was introduced into Mexico where it underwent further domestication and selection which assisted in the production of some of the more recognizable characters of present-day tomatoes before a wider distribution took place (Jenkins 1948).

There are two main types of tomato plant, the determinate (or bush) and indeterminate, trained as a single stem, or 'vine', by frequent removal of the side shoots as they emerge from the leaf axils. Several authors have discussed the mode of growth in the two tomato types and for a comprehensive and detailed account the reader is referred to the description by Luckwill (1943).

The determinate or 'bush' tomato cultivars are widely grown in the world and are used for the majority of the field crops grown for the fresh market or processing, such as canning, *purée*, soup, juice or ketchup. Cultivars suitable for modern mechanical harvesting systems have been developed by plant breeders. These processing cultivars, which are often mechanically harvested, exhibit many important characters, such as a large number of fruits set at any one time which subsequently ripen over a short period, longevity of ripe fruit on the plant, the so-called 'jointless' character (which enables the fruit to detach from the calyx where it joins the fruit rather than the abscission layer or 'knuckle'), and a relatively high dry matter content.

Similarly, the modern indeterminate cultivars have been largely produced by plant breeders for the greenhouse industry, where there is a relatively high capital investment per unit area, and they are also used in some field production systems where high quality fresh fruits are required, usually for salad or 'slicing', and where there is adequate hand labour for training the plants and picking the fruit over a prolonged marketing period. The indeterminate types are also popular among private or home gardeners, especially in Europe and North America, although there is a current trend by seed houses in North America to make improved determinate cultivars more available to the home gardener.

Many other characters have been included in breeding programmes of both the determinate and indeterminate types of tomato cultivars, probably the most important of which are resistance to specific pests and pathogens with improvement in fruit qualities, such as shape, size and even-ripening; this last character reduces the incidence of green shoulders of the ripening fruit.

Plant breeders have also successfully combined desirable characters in F_1 hybrids produced for specific environments and market requirements for both extensive field and intensive greenhouse production.

Cultivar description of tomato

Season	early, mid-season or maincrop
Use	processing (e.g. *purée*, juice or canning), fresh market or storing
Plant habit	determinate or indeterminate vigour: general height of plant, relatively short or tall
Leaf characters	cut leaf or potato leaf

| Stem characters | hairiness |
| | colour: green, blue |

Flower characters — colour, type of style (see Fig. 12.4)
relative size

Fruit characters — (some fruit characters are detectable before ripening)
colour
relative size
shape: e.g. oblate, globe, square, pear (see Fig. 12.1) cylindrical or intermediate shape between any of these
external features: e.g. smooth or ribbed

Resistance to specific diseases: e.g. *Cladosporium fulvum*, *Fusarium oxysporum* or *Verticillium dahliae*

Resistance to specific pests: e.g. *Meloidogyne* spp.

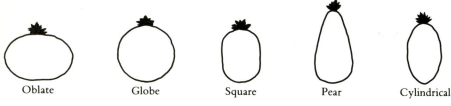

| Oblate | Globe | Square | Pear | Cylindrical |

Fig. 12.1   Basic tomato fruit shapes.

Agronomy

The plants are usually raised in nursery beds and planted out bare-rooted into the field or seedlings are raised and potted on individually into suitable pots or containers before they are planted into their final quarters.

There is relatively little direct drilling or sowing into the field anywhere in the world. The practice of raising young plants as a separate operation enables the grower to achieve a high degree of plant uniformity and also provides a check against early pests and diseases while the plants are still in the nursery. The use of nursery beds also allows a relatively early start to the crop under plastic or glass protection before planting out. Figure 12.2 shows a typical tomato seedling nursery in Asia.

The final stand and density of plants in the field depends on whether the cultivar is a trained or bush type. The traditional spacing for staked or 'trellis' plants is 30–60 cm within the rows and 90–120 cm between the rows.

Bush cultivars are usually spaced at 45–120 cm within the rows and 90–150 cm between the rows. But the final decision on plant population will depend on the cultivar, irrigation system and whether or not a support system is used on which to train the plants. In greenhouse production the plant population is approximately 3000 plants per hectare.

There are approximately 400 seeds per gram for most cultivars, although those used by the greenhouse industry tend to be larger seeded, approximately 300 seeds per gram.

Fig. 12.2   Nursery production of tomato seedlings.

Sowing rate

For transplants, 60 g of seeds are drilled per 250 m² of seed bed which will provide sufficient plants for 1 ha. When direct drilled, the sowing rate is between 500 and 1000 g per hectare.

Nutrition

The pH should be maintained at 6.5; appropriate quantities of liming material must be added during preparation if the pH is lower than this. The disorder commonly called 'blossom-end rot' or 'black spot' (see Fig. 12.3) is the result of calcium deficiency, although even when there is sufficient calcium available from the soil or substrate, the condition can arise when the soil moisture level has been allowed to become low for about two days or longer.

The nutrients applied during field preparation are calculated according to the existing status of the soil. The following amounts are applied on soils of relatively low fertility: N 75–100 kg per hectare; P_2O_5 150–200 kg per hectare; K_2O 150–200 kg per hectare.

For nutrition and irrigation under protected cropping the reader is referred to the *Tomato Growers' Handbook* (Kingham 1973).

A relatively high level of available phosphorus is important for a satisfactory tomato seed yield and quality when plants are grown in greenhouses (George *et al.* 1980). But while a high nitrogen level has been shown to pro-

Fig. 12.3   Tomato blossom and rot.

duce a high quality seed, under field conditions the nitrogen status should be about half the potassium status in order to maintain a good balance between inflorescences and vegetative growth.

Flowering

The tomato is normally grown commercially as an annual crop, although under continued ideal conditions it tends to be a relatively short-lived perennial.

The plant cannot survive temperatures at or below freezing and even prolonged periods of less than 10 °C will reduce crop productivity significantly. It can survive relatively high temperatures and continues to grow very successfully up to about 25 °C; at higher temperatures than this there can be an adverse effect on growth rate and fruit yield even though the plant appears to be in a satisfactory vegetative condition.

The flower initiation is not dependent on photoperiod, although there is a tendency for flower truss initials to abort in the relatively low winter light levels experienced, for example, in the greenhouses of Northern Europe. The plant responds well to high light levels.

Pollination

The tomato is generally considered to be a self-pollinating crop. Its flower structure favours complete self-pollination without the need of insects or other natural pollinating agents; most cultivars, especially those developed for greenhouse crop production, have continued to maintain this trait. There is likely to be some cross-pollination, however, which can occur for one of several reasons. Firstly, there is always the possibility of an appropriate insect introducing foreign pollen from outside the crop and this may be increased when flowers have been emasculated in preparation for F_1 hybrid seed production.

Another factor is the presence of pollinating insects capable of cross-pollination. These insects are much more likely to be encountered when tomatoes are produced in or near their original habitat of South America (Richardson and Alvarez 1957). The possibility of insects with similar capabilities being encountered in other production areas cannot be ruled out.

Thirdly, despite the trend of plant breeders and seed producers to select towards complete self-pollination, there are some cultivars available whose flowers have long styles which tend to favour cross-pollination. This factor and its effect on natural cross-pollination in tomatoes was first discussed by Lesley (1924). The majority of modern tomato cultivars have short-styled flowers. Diagrams of long- and short-styled tomato flowers are shown in Fig. 12.4.

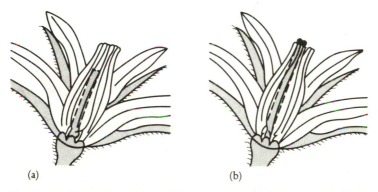

(a)   (b)

Fig. 12.4   Short (a) and long (b) styled tomato flower types (after Rick 1978).

There is a noticeable reduction in fertilization when ambient temperatures reach about 42 °C. This occurs in tropical field crops when high temperatures persist for prolonged periods. The poor setting resulting from high temperatures is not only attributed to a decrease in pollen fertility, but also to physiological factors affecting fertilization. This has not yet been explained by research workers but has been demonstrated by pollinating flowers of plants growing under high temperatures with good quality viable pollen from elsewhere and the subsequent seed set is still relatively poor. Under greenhouse conditions it is considered necessary to agitate the plants daily to assist the transfer of pollen from the anthers to the stigmas. A useful confirmation that pollination is taking place at a satisfactory rate is when only two successive flowers can be seen open on a truss at any one time. Preceding flowers on the truss would have been pollinated and succeeding flowers will still be in bud.

The period from fertilization to ripe fruit will depend on the cultivar and environment. This period can be between forty and sixty days but the usual time expected in most tomato production areas is about forty-five days.

Isolation

The minimum isolation distance between different cultivars of tomatoes for seed production is relatively short. This is because of the crop's high level of self-pollination. Most countries specify a distance of between 30 and 200 m, the main consideration being that the distance should be sufficient to avoid admixture at harvesting time.

The distances between two parental lines grown for the production of a specific F_1 hybrid need be no more than 2 m, and less than this if there is a physical barrier such as under greenhouse conditions.

Previous cropping

The area should not have grown tomatoes in the previous two years, but if soil sterilization using heat or fumigation is done after each crop with a chemical soil sterilant, such as methyl bromide, then successive crops can be grown in the same area. Regular partial soil sterilization or soil fumigation is practised in greenhouses where rotation is less practicable and the overall investment is much higher than field production. The authorities in some countries insist that no solanaceous crop should have been grown for at least the previous three years. This is more from the point of view of preventing pest and disease build-up than reducing the possibility of seedlings from previous crops.

Tomato roguing stages and main characters to be observed

Plants showing characters adverse to the type to be removed.

1. *Before flowering*   Note desirable characters: growth habit and foliage typical of the cultivar; leaf characters, general habit, width. Observe if specific diseases present.

2. *Early flowering and first fruit immature*   General plant habit and characters as defined for 'before flowering'. Immature fruit according to the cultivar description, e.g. greenback or greenback-free.

3. *Fruiting*   Yield characters, fruit quality, shape, colour when ripe, relative size.

Production of F_1 hybrid tomato seed

· The majority of F_1 hybrid tomato seeds used in the world are produced under field conditions. However, some hybrid seed of indeterminate cultivars is produced in greenhouses, especially in The Netherlands and Great Britain, for the specialized protected cropping industries of Europe and elsewhere.

The production of hybrid tomato seed involves the maintenance and growing on of two separate lines of plants; i.e. the male and female parents. The male is used as a source of pollen and the female receives the pollen and subsequently produces the seed-bearing fruits. Thus, in practice, two separate batches of plants are grown from seed of two specific lines which have normally been produced and maintained by the breeder who develops or maintains the hybrid. The seed producer grows the two parents and does the hybridization in accordance with the original breeder's instructions. For example, using one appropriate parent (male) for the supply of pollen and the other (female) which receives the pollen and subsequently produces the fruit containing the F_1 hybrid seed.

It is normal practice to sow the male parent up to three weeks earlier than the female to ensure an adequate supply of pollen for fertilization of the maximum number of flowers on the female plants. Local observations of the time from sowing until start of flowering on each parent line will provide evidence for sowing dates in future years.

The ratio of male line plants to females will depend on the flowering habits of the individual lines, but a ratio of approximately 1 : 5 male to female is a general guideline.

Before hybridization commences all plants in both lines to be used in the programme should be finally checked for trueness to type; any off-types, undesirable or suspect plants should be removed.

The female parent plants of indeterminate types are normally supported by an appropriate system according to local conditions and custom. Satisfactory support is especially important when the operation is done under protection. Potential total plant and fruit weight must be taken into account in the design of the support system. It is usually considered unnecessary to support the male parent, except possibly when indeterminate male types are grown in protective structures or when a supply of pollen is likely to be needed over a long period.

The flowers on the female line must first be emasculated during their late bud stage in preparation for cross-pollination. This involves removal of their anthers and is done as a separate operation prior to applying pollen to their stigmas. The anthers which are fused in a cone are cleanly removed by a pull, using forceps. This operation requires practice and new workers should be given the opportunity to develop and practise their skill before the hybridizing season starts. It may be necessary to dissect off a single anther from the fused anther cone before being able to pull off the remainder in a single pull. It is extremely important not to damage the stigma at all during the operation otherwise it will not be receptive to pollen and fertilization will not occur.

The timing of emasculation in relation to the opening of individual flowers is related to the speed at which flowers develop. Experience under local climatic and environmental conditions indicates how long in advance of 'normal' opening the flowers should be emasculated. For example in Californian field production, flowers are emasculated early in the morning (from about 06.30) on the same day as pollination; whereas in North European greenhouse hybrid seed production, flowers are emasculated up to two days in advance of pollination.

One of the most useful assets in hybrid seed production is to have a female parent which is male sterile. Male sterile lines for hybrid tomato seed production exist to some extent but it is difficult to obtain information from the seed houses as to the number of hybrid cultivars being produced using male sterility. However, there is no need for emasculation if male sterile female parents are in use. But the task of hand pollination still has to be done.

Pollen is collected from the male line either by detaching individual flowers or by mechanical means such as a small vibrator. This apparatus which is often referred to as an 'electric bee' is a hand-held battery instrument. It is used to agitate individual open flowers which release pollen on to a small receptive plate, disc or tube attached to the instrument.

The pollen which has been collected is applied with a fine brush to each emasculated (or male sterile) female flower. In some Asian countries, specializing in the large-scale field production of hybrid tomato seed, workers transfer the pollen from the collecting receptacle to the stigmas by finger. Each pollinated flower is marked with a coloured tag and some seed producers remove about half the calyx from pollinated flowers immediately after pollinating; this later provides a clear confirmation of fruits containing hybrid seed at harvest time.

In large-scale field production of hybrid tomato seed it is customary for the hybridizing gang to divide into two sections with each section starting at either end of the rows; working in pairs with one either side of the row, they progress towards the centre of the plot. Under less extensive greenhouse conditions individual workers usually have plots or rows for which they are individually responsible throughout the pollinating season.

Under field conditions in arid areas the whole operation is usually done three times a week on alternate working days. But in seed production under protection in temperate areas the plants may be looked over daily to emasculate flower buds at the correct stage. Any unhybridized flowers are removed or remain on the plant unmarked.

Every care must be taken to mark clearly and label separately plants of either parent if they are to be maintained by natural selfing for production of seed for future stock.

Tomato picking and seed extraction

Seeds are extracted from ripe fruits which have either been hand picked into containers or collected by a single mechanized harvester which removes all the fruit from the crop in a single operation. The method of picking adopted locally depends on the scale of the operation and level of technology available in each seed production area.

Fruit resulting from hybridization is always hand picked from the female lines whether it is produced in the field or under protection. This ensures that only the fruits which have been cross-pollinated are included. They are identified from either their label or calyx marking. This check, or confirmation, at the time of harvesting fruit is extremely important when hybrid seed is being produced in order to exclude fruit which contains seed produced by natural selfing of the female line.

Seed extraction of small quantities of fruit

The ripe fruits are cut equatorially and the seeds with the gelatinous material surrounding them are squeezed or spooned out into containers. During this process the main fruit walls, pulp, skin and other debris are excluded. The seeds and the gelatinous material are then separated by one of the processes described later.

Seed extraction of large quantities of fruit

The extraction of seeds from large quantities of tomato fruit can be divided into two main types of operation. Firstly, when the seed extraction is the only commercial operation involved, and secondly, when the seed is extracted as an additional product during the processing of tomatoes for *purée* or juice.

Commercial seed extraction

Here the word 'commercial' is referring to the main purpose of the operation and not to the seed purity or class, as this method is frequently used for large-scale production of basic and certified seed as well as commercial seed.

This system can be completely mechanized in that the fruit is harvested automatically before passing into a crusher. Alternatively, the fruit is hand picked and either put directly into the crusher as it slowly passes through the field at a speed in keeping with the rate of hand picking, or the picked fruit is transported to a stationary crusher in a seed cleaning yard or special area.

The crusher squashes or crushes the fruit and the resulting mixture of gelatinous seed, juice and fruit residues is passed through a screen to separate off the gelatinous seed from the bulk of the remaining material. The crushed material is usually passed into a revolving cylindrical screen which allows the seeds and juice to pass through the mesh, while the fruit debris passes through the cylindrical screen to drop in the field. Alternatively, the debris is collected separately for later disposal if the operation is stationary. The juice and seed mixture is collected in separate containers. These machines are usually purpose built by, or for, the seed company or organization. The field operation of this type of machine is shown in Fig. 12.5. The separated seed is then finally extracted from the gelatinous material and other materials by a separation process described later.

Combined juice and seed extraction

This system of seed extraction is done in co-operation with a tomato processing factory. The factory line is generally organized to produce *purée* or juice processed for domestic use. During the operation the *purée* or juice is separated from the relatively dry residual mixture of seeds, pulp and skins. Special lines of apparatus are used in processing plants which intend to secure the seed in this way. One important feature is that the seeds are not subjected to the high temperatures used during tomato processing, and a purpose-designed modification of the normal processing plant is frequently referred to as a 'cold take-off'. The apparatus is normally made and sold by the specialist manufacturers of industrial tomato processing equipment.

Fig. 12.5   Tomato fruit crusher and seed extractor.

Close liaison must be maintained between the seed producing and the to-mato processing organizations. One important feature is the need to have large batches of fruit of the same cultivar going through the factory and a thorough cleaning technique to avoid admixture of seeds from different cul-tivars. The system is used for the production of commercial quality seed for large-scale industrial tomato crops. It should not be used where seed of high genetical quality is required for further multiplication. Seed which has been extracted by this process has usually been separated from the mucilage during the industrial process and all that is required is washing to separate the seed from the other debris.

When the tomato seed has been extracted from the fruit by one of the fore-going operations or systems, it is then usually necessary to separate the seed by a further wet-upgrading method referred to as 'separation' before it can be finally washed and dried. The method of separation will depend on quan-tity of seed to be processed, possible need to control a specific seed-borne pathogen or virus, and the temperature of the local environment.

Separation by fermentation

The pulp containing the extracted tomato seed is left to ferment for up to three days at about 20–35 °C. But the rate of fermentation will depend on the ambient temperature and may even take up to five days. In warmer areas such as California the fermentation process is usually completed within twenty-four hours. Frequent inspections will determine when the seed's gel-atinous coating has broken down. The mixture must be stirred several times a day to maintain a uniform rate of fermentation in the container and to avoid discoloration of the seed. It is usually necessary to cover the containers with muslin to reduce frit or fruit fly activity. The fermentation time must not be extended beyond that required for breaking down the mucilage or else the subsequent seed quality will be affected by premature germination.

There are claims that the fermentation process controls seed-borne bacterial canker of tomato *Corynebacterium michiganense* (E. F. Smith) Jenson, but the measure of success will depend on the time of fermentation and the temperature reached. Doolittle (1948) suggested that a fermentation temperature of 21.1 °C should be maintained for ninety-six hours.

Separation with sodium carbonate

This method is relatively safe and can be used for small quantities of seed in cooler temperate areas where the fermentation method is not used. The pulp containing the extracted tomato seed is mixed with an equal volume of a 10 per cent solution of sodium carbonate (washing soda). The mixture is left for up to two days at room temperature after which time the seed is washed out in a sieve and subsequently dried. The sodium carbonate method of extraction tends to darken the testa of the seed and is therefore not normally used for commercial seed-lots, but it is used by plant breeders and other workers who are involved in maintaining breeding material and inbred parent lines.

Separation with hydrochloric acid

This method is often favoured by large commercial producers as it produces a very bright clean seed sample. The actual dilution rate and duration of treatment used by the major commercial seed producers is usually a closely guarded secret. The hydrochloric acid treatment is often combined with the later stages of fermentation. However, producers of relatively small quantities of tomato seed find that 567 ml of concentrated hydrochloric acid stirred into 10 litres of seed and pulp mixture and left for half an hour is successful. It is very important that the acid is added to the water and pulp, and *not* the water and pulp to the acid, otherwise a dangerous effervescence will occur. All workers handling the concentrated and diluted acid solutions must wear appropriate face shields and protective clothing.

The hydrochloric acid is damaging to seeds of the compact ('shortjointed') cultivars which are used by some specialized growers in Britain and therefore it should not be used when producing seed of these types.

Control of seed-borne tomato mosaic virus (TMV)

The first principle of minimizing the transmission of TMV via the tomato seed, in addition to the appropriate isolation mixtures discussed later, is to ensure that fruits for seed evaluation are only harvested from plants which do not show symptoms of infection. Where a relatively small number of plants is involved (which is especially possible where seeds of the determinate types are produced in greenhouses) fruit for seed extraction should be harvested only from the lower trusses of late-infected plants if healthy ones are not available (Broadbent 1976).

There is evidence that both the sodium carbonate and the hydrochloric acid extraction methods will inactivate tomato mosaic virus transmitted on the testa (Alexander 1962; Nitzany 1960). Another treatment, involving the use of trisodium orthophosphate, is used as an extra safeguard to inactivate high value tomato seed used by the greenhouse industry. It is a separate treatment which is done immediately following extraction but before the seeds are dried. The seeds which have been extracted by one of the foregoing methods are soaked in a 10 per cent trisodium orthophosphate solution for thirty minutes and are then immediately rinsed and dried.

Tomato seed washing

Tomato seeds which have been extracted by fermentation or acid treatment are washed immediately the extraction time has been completed. This can be done on a small scale by washing in a series of sieves.

Large quantities of seed extracted from field crops are usually washed in long water troughs with a fall of 1 in 50 (see Fig. 12.6).

The trough has riffles at intervals and works on the principle that the seed is denser and sinks while the other fruit debris is floated off or goes through the water trough in suspension. The skill of doing this operation comes with experience and it is always a safeguard to have a suitably sized screen over the waste-water drain in case of mishaps.

Tomato seed drying

This operation should commence immediately the washing process has finished. Large-scale extraction units may spin dry the seeds while they are in a stout cloth bag to remove excess water before the main drying. A typical commercial size spin drier used for tomato seeds is illustrated in Fig. 12.7.

In warm dry climates the seeds can be spread on hessian or other suitable mats and sun-dried, in which case the seeds are brushed to turn them occasionally. Large quantities of seeds are dried in rotary paddle hot air batch driers as described in Chapter 4.

Tomato seed yield

Figures quoted for tomato seed yield vary significantly from one production area to another and it is not possible to quote an average seed yield. The differences will vary according to several important features which include:

Type of plant	determinate or indeterminate
Number of fruit trusses per plant	this can vary extensively for greenhouse cultivars, according to number of fruiting trusses

Fig. 12.6 Washing extracted tomato seeds in water troughs.

Fig. 12.7 Spin drier for extracting excess moisture from wet extracted seeds.

Plant density or popu-   this will be dictated, for example, by training
   lation per unit area   system and irrigation system.

Different reports quote two main ways of estimating tomato seed yield:

(a) according to weight of fruit; (b) according to unit area of plants.

In greenhouse production 1 kg of fruit will produce approximately 4 g of seed (approximately 1200 seeds). In field production one rule of thumb is that the seed weight is 1 per cent of the fruit weight, e.g. 1 tonne of fruit will produce 1 kg of seed. The expected tomato seed yield in the USA is between 250 and 400 kg of seed per hectare. Workers in Africa report yields from 10 to 50 kg per hectare.

The greenhouse tomato cultivars with a 1000 grain weight of 3.3 g tend to have larger seeds than the field or determinate types. The field or determinate types have a 1000 grain weight of 2.5 g.

The multiplication factor is not a very reliable method of calculating seed yield in tomatoes because of the points outlined above and the fact that plants are raised in nursery beds before transplanting. However, a multiplication factor of 200 can be used for the indeterminate types, while the determinate types under field conditions have a multiplication factor of 50.

The main seed-borne tomato pathogens with common names of the diseases they cause

Pathogens	Common names
Alternaria solani Soraner	Early blight, *Alternaria* blight
Cladosporium fulvum Cooke syn. *Fulvia fulva* (Cooke) Ciferri	Leaf mould
Fusarium oxysporum Schlecht ex Fr.	*Fusarium* wilt
Glomerella cingulata (Stonem.) Spauld and Schrenk	Anthracnose, ripe rot
Phoma destructiva Plowr.	Fruit and stem rot
Phytophthora nicotianae B. de Haan var. syn. *parasitica* (Dastur) Waterh.	Buck-eye rot
Rhizoctonia solani Kühn	Damping-off, collar rot
Verticillium dahliae Kleb.	*Verticillium* wilt, sleepy wilt
Corynebacterium michiganense (E. F. Smith) Jenson	Bacterial canker, Grand Rapids disease
Viruses	Tobacco mosaic virus
	Tomato bunchy top virus, Mosaic
	Potato spindle tuber viroid

EGGPLANT: *Solanum melongena* L.

Origins and types

This species has several widely used common names including eggplant, aubergine and brinjal. The eggplant originates from India where a wide range of wild species and land races occur and is now generally grown as a vegetable throughout the tropical, sub-tropical and warm temperate areas of the

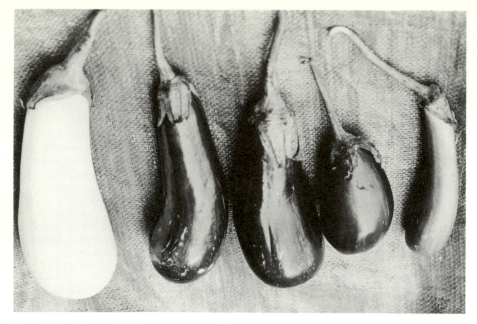

Fig. 12.8   Range of fruit types in eggplant.

world. In recent years it has increased in importance as a protected crop in Northern Europe.

The cultivars of *S. melongena* L. display a wide range of fruit shapes and colours, ranging from white, yellow, green, through degrees of purple pigmentation to almost black. The range of fruit shapes and extreme colours is shown in Fig. 12.8.

Eggplant is generally grown as an annual although it is a weak perennial. The plants usually develop some spines which are especially noticeable on older plants, but the degree of spininess can also depend on the cultivar.

Cultivar description of eggplant

Season	early or maincrop, approximate number of days to first mature fruit
Plant habit	height – relative height and vigour of plant
Leaf characters	shape, size and pose
Stem characters	colour, extent and degree of downiness, spines, size of spines and degree of spininess on the mature plant
Flower characters	number of flowers on inflorescence, borne singly or in groups
Fruit characters	colour while ripening and when mature – white, green, yellow, purple to black and bi-colour if specific to cultivar when ripe

shape and length – cylindrical, globular or bulbous
size – position of mature fruit on plant in relation to
   soil level
exposure of fruit

Calyx characters   shape, colour and size relative to fruit

Resistance to specific diseases e.g. *Phomopsis vexans*

Resistance to specific pests e.g. *Meloidogyne* sp.

Flowering

The eggplant does not have a specific daylength requirement for flower initiation, but requires a higher minimum temperature than tomato. Optimum
day temperatures for satisfactory growth and fruit production are around
25–35 °C with lower average night temperatures of 20–27 °C. The eggplant
is less tolerant to low temperatures than tomato and cannot survive frost. If
subjected of temperatures at the lower end of the temperature range outlined
above the growth and development are very slow. However, eggplant with-
stands humid tropical conditions better than tomato.

Pollination

The flowers of eggplant are normally self-pollinated although some cross-
pollination occurs when sufficient bees are attracted to the flowers. Choudhury (1971) found that the amount of cross-pollination under Indian
conditions ranged from 6 to 20 per cent when plots are less than 50 m apart,
but no cross-pollination at distances greater than this.
   There are F_1 hybrid cultivars produced by specialist seed companies using
male sterile lines as the female parent, or more commonly by emasculation
followed by hand pollination. When male fertile female parents are used in
the cross they are bud pollinated.
   The F_1 hybrids of eggplant show some heterosis but the greatest benefits
claimed by commercial seed producers are advantages such as earliness and
tolerance to pathogens, for example tobacco mosaic virus and *Phomopsis vexans* (Sacc and Syd) Harter (fruit rot).

Plant raising

Eggplants are normally grown from a transplanted crop which has been
raised in nursery beds. In some areas seedlings are raised in containers or seed
beds before pricking-off into nursery beds when the seedlings show their first
true leaves.
   When seedlings are to be pricked out the sowing rate is 60 g to 250 m^2
which provides sufficient plants to plant out 0.5 ha.

Spacing

The optimum spacing of plants in their final positions is from 46 to 76 cm between plants in the rows and 60 to 120 m between the rows, depending on the vigour of the cultivar and any irrigation system.

Rotation

Eggplants are susceptible to many of the soil-borne pests and diseases associated with other members of Solanaceae and the rotations must take this into account. Generally a period of four years should elapse between successive eggplant crops or eggplants and other genera in Solanaceae.

Nutrition

The soil pH should be approximately 6.5 and if lower than this sufficient lime should be added to bring it up to about 6.5.

Eggplants may be grown on raised beds in the humid tropics especially where occasional or regular heavy rainfall is expected, although large-scale seed production from plants grown in raised beds is unlikely to prove economic.

The crop responds to relatively high nutrient levels. Base dressings of 80–120 kg nitrogen per hectare, 80–100 kg P_2O_5 per hectare, and 80–100 kg K_2O per hectare are given according to soil nutrient status.

Nutritional studies made in India by Seth and Dhaudar (1970) have shown that the yield of eggplant seed increased at the relatively high levels of 70 kg/ha of nitrogen compared with lower levels of nitrogen of 30 and 50 kg/ha. Although the N, P and K levels used by Seth and Dhaudar were lower than the general recommendations given earlier in this section, they demonstrate the value of nitrogen in increasing yields of eggplant seed. The nutrient levels used by Seth and Dhaudar are probably closer to the nutrient regimes used by small-scale farmers and seed producers in developing countries, whereas the higher recommendations are more in line with the large-scale farmers producing eggplant seed under contract in the more developed areas.

The importance of nitrogen on accelerating flower formation in eggplant has been suggested by Eguchi *et al*. (1958) working in Japan.

Eggplant roguing stages and main characters to be considered

1. *Before flowering*   Note desirable characters – plant habit, and foliage typical of the cultivar. Pose, shape and relative size of leaves.

2. *Early flowering and first fruit developing*   General plant habit, vigour and characters as defined for stage 1. Degree of spininess.

3. *Fruiting*   Yield characters, fruit quality, shape, size, colour and patterning (if appropriate) when ripe. Internal fruit colour (basic seed).

Harvesting

In order to ensure that seed development is complete, the fruits are usually hand picked at a later, or riper stage than for the market crop. Some seed producers leave the fruits on the plant until the abscission layer just behind the calyx is fully developed.

If hybrid seed is being produced care must be taken to exclude fruit not resulting from hybridization and in this case it is better to pick fruit before it is likely to drop off the plant.

Seed extraction

There are two basic methods used for the extraction of eggplant seeds, a wet and a dry extraction process. There is a general tendency to favour the wet extraction especially in large-scale seed production. The dry process is still favoured where relatively small amounts of seeds are produced.

Wet extraction

For the primary extraction of eggplant seeds, the fruits are crushed and seeds are separated from the remainder of the fruit pulp and debris as described for tomato seed extraction. But because eggplant fruit pulp is relatively dry, it is necessary to add extra water during and after crushing in order to improve the separation of the seeds.

The seed is extracted from the debris by the same process as tomato, but it is sometimes necessary to spray clear water into the separation screen cylinder to obtain the maximum seed yield. The extra water tends to wash the seeds free from the pulp.

Dry seed extraction

In some countries the over-ripe eggplant fruits are dried in the sun until they shrivel. In the purple and purple-black fruited cultivars the drying out is accompanied by a fading of the skin colour to a coppery brown colour. The fruits are then hand beaten and the dried seeds hand extracted.

This method is time consuming and laborious, but is used in some countries for production of relatively small seed-lots when ripe fruits are accumulated over several weeks and hand labour is available for the final extraction (Fig. 12.9).

Seed yield and 1000 grain weight

A satisfactory seed yield is approximately 150 kg/ha, but high yielding cultivars grown in areas of good crop husbandry can yield up to 200 kg/ha.

The 1000 grain weight is approximately 5 g but the smaller seeded cultivars have a 1000 grain weight of *c.* 4 g.

Fig. 12.9 Extracting eggplant seeds by hand from dried fruit.

The main seed-borne eggplant pathogens with common names of the diseases they cause

Pathogens	Common names
Alternaria alternata (Fr.) Keissler	Leaf spot and fruit rot
Colletotrichum melongena Lobik	Anthracnose
Fusarium oxysporum Schlecht ex Fr.	*Fusarium* wilt
Phomopsis vexans (Sacc. and Syd.) Harter syn.	Fruit rot
Diaporthe vexans Gratz.	

Pathogens	Common names
Rhizoctonia solani Kühn	Damping-off
Sclerotinia sclerotiorum (Lib.) de Bary	
Verticillium albo-atrum Reinke and Berth.	*Verticillium* wilt
Verticillium dahliae Kleb.	
Virus	Eggplant mosaic virus

PEPPERS

The peppers which are grown as vegetables are *Capsicum* spp., not to be confused with *Piper nigrum* L. which is also commonly called 'pepper', the fruits of which are used as a spice or condiment.

Some authorities, e.g. Heiser and Smith (1953) distinguish more than two *Capsicum* spp., cultivated as vegetables, while Bailey (1948) recognized only *C. frutescens* L. (syn. *C. annuum* L.) but described five botanical varieties. The present author accepts the proposal of Purseglove (1974) that these are all forms of either *C. annuum* or *C. frutescens*.

Capsicum annuum L. is usually an annual with flowers and fruits borne singly. Cultivars of this species include the chillis, red, green, yellow, sweet peppers and paprika.

Capsicum frutescens L. is a shrubby perennial with several flowers on each inflorescence; cultivated types include the bird chillis, cherry capsicum and the cluster pepper, although it is not clear whether cluster pepper is a type derived from *C. annuum* or *C. frutescens*.

Capsicum peppers originated in South America and spread into the New World tropics before subsequent introduction into Asia and Africa. They are now widely grown throughout the tropics, sub-tropics and warmer temperate regions of the world. Peppers are not frost tolerant. In recent years the larger fruited sweet and condiment forms have been increasingly grown as a greenhouse crop during winter or summer in Central and Northern Europe.

Cultivar description of pepper

Season	early, mid-season. Duration of cropping period
Type	sweet (mild) or pungent (hot)
Use	fresh market; processing, pickle, sauce, dehydration
Plant habit	upright, branching
Leaf characters	flowering over long or short period. Distinctive flower characters, e.g. colour
Fruit characters	position of mature fruit on plant. Pose of fruit on plant, e.g. horizontal or upright
	relative size
	external fruit characters (Fig. 12.10):

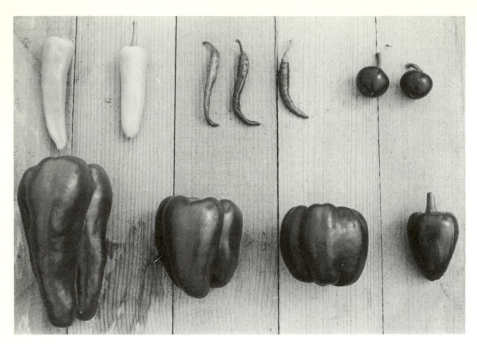

Fig. 12.10   Fruits from different pepper cultivars showing diversity of size and
shape.

shape – square (block), heart, cylindrical, round,
'cherry', 'tomato', long thin
colour (when ripe – yellow, orange-red, dark red,
black or bi-colour
internal fruit characters:
number of lobes
degree of pungency
Resistance to specific diseases

Flower formation and seed set

The physiological requirement for flower initiation in *Capsicum annuum* has
been studied by several workers since Cochran's (1932) earlier work. Rylski
(1972) examined the interaction of temperature with flower development and
confirmed Cochran's earlier findings that the species comes into flower
sooner in short days than long days. Rylski also showed that anthesis
commences earlier at relatively high night temperatures and reported a decrease in time to anthesis of first flower from sixty-seven to fifty-one days
when subjected to a night temperature of 10 °C compared with 25 °C respectively; the number of leaves prior to flowering was inversely proportional
to the night temperature, but this is of less importance in seed production
whereas choice of site in relation to ambient night temperatures can be extremely useful especially in areas with relatively short seasons.

Effect of temperature on seed set

Peppers are able to produce parthenocarpic fruits and fruits containing very few seeds when the night temperature is relatively low. The effect of low temperature is greater before anthesis than afterwards. While this is an advantage for the production of a market fruit crop, it is not helpful in seed production. Rylski (1973) investigated the effect of night temperature on shape and size of sweet pepper fruit. Data shown in Table 12.1 demonstrate the temperature effect on seed yield.

Seed yield can be considerably less for pollinated flowers under low temperature conditions than those used in Rylski's experiments. Figure 12.11 shows the effect of low temperatures on seed yield up to fruit ripening.

Relatively low temperatures also affect fruit size and shape. Any early detection of adverse fruit shape should be investigated to ensure that pollination

Table 12.1   Effect of night temperature before and after anthesis of pollinated pepper fruits (data from Rylski 1973)

Night temperature before anthesis (°C)	Night temperature after anthesis (°C)	Number of seeds per fruit
18–20	8–10	251
18–20	18–20	252
8–10	8–10	235
8–10	18–20	204

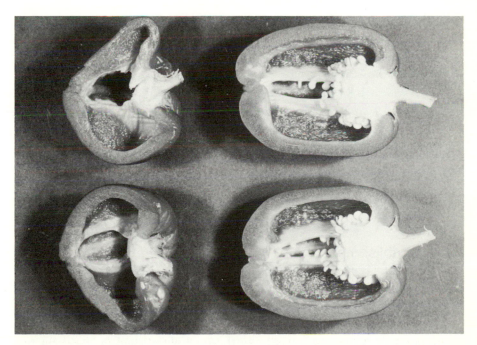

Fig. 12.11   Longitudinal sections of the same cultivar of pepper fruits showing satisfactory and poor seed set.

is satisfactory. Other causes of adverse fruit shape, not necessarily associated with lower seed set, may be genetically controlled.

Pollination

Capsicum annuum and *C. frutescens* are generally self-pollinated, but some cross-pollination can occur between and within cultivars of the two species. Murthy and Murthy (1962) reported up to 68 per cent cross-pollination in India. The high degree of cross-pollination experienced occasionally is due not only to large populations of pollinating insects, but also to dehiscence of the anthers up to two or three days after individual flowers open. Prior to dehiscence the stigmas are receptive to pollen transmitted from other plants.

It should be assumed that all the types of cultivated peppers are cross-compatible when producing seed rather than attempting to subdivide the genus into several species with different degrees of cross-compatibility.

Nutrition

The pH of the soil should be between 6.0 and 6.5, and if lower than this a liming material should be added.

Nutrient regimes with a relatively high level of nitrogen should not be used as they tend to delay fruit maturity. A general basic nutrient application during preparation is 115 kg nitrogen per hectare, 200 kg P_2O_5 per hectare, 200 kg K_2O per hectare.

However, the quantities of individual fertilizers applied should be modified according to the soil's nutrient status. Up to two-thirds of the nitrogen can be applied as top dressings once flowering and fruit setting have commenced, especially in the humid tropics.

Irrigation

If the crop is to be grown in an area which is not rain-fed, an adequate supply of irrigation water must be available because the capsicums are very prone to flower-drop. This and the dropping off of undeveloped fruits occur during times of water stress, more so with the larger fruited types.

Field establishment

Peppers are either grown *in situ* or plants are raised in nursery beds and planted out as described for tomatoes. Raising the plants in nurseries ensures that the seedlings can be protected from late frosts, heavy rain or wind.

Final plant density depends on both the vigour of the cultivar and the irrigation system to be used. Generally the final plant stand is 30–60 cm between plants in the rows and 45–90 cm between the rows.

Pepper roguing stages and main characters to be observed

1. *Before flowering* Note desirable characters – growth habit, vigour and foliage typical of the cultivar. Leaf characters. Observe if any specific seed-borne disease present.

2. *Early flowering and first fruit immature* General plant habit and characters checked at stage 1. Observe if any specific seed-borne disease present.

3. *Mature fruit* General plant habit and characters checked at stage 2. Fruit colour when ripe. Fruit size, shape and length. Observe if any specific seed-borne disease present.

Basic seed production

Crops for basic seed are rogued, or plants are selected, according to the criteria listed under cultivar description and roguing stages. Increased pungency can be inadvertently introduced into sweet or mildly pungent types by pollen contamination from more pungent types and it is therefore important to check that the absence of, or degree of pungency is according to the cultivar type. This can be achieved by examining and tasting a small piece of the fruit's placenta-wall tissue from each selected plant. This process is not necessary for the hot chilli types, but it is important that the stock of a mild type does not increase in pungency. The internal fruit characters to be examined for basic seed also include thickness of fruit wall and internal fruit flesh.

Seed extraction

There are two basic methods of pepper seed extraction, a wet and a dry. It is not possible to dry completely the large fleshy fruited sweet types, and they are therefore usually wet extracted, without fermentation. The crushing and wet extraction process is similar to that described for eggplant.

The small fruited pungent types can be successfully dried before seed is extracted in the sun or put into batch driers until shrivelled and dry. In areas where there is sufficient hand labour the dried fruits are hand flailed to extract the seeds but elsewhere they are put through a thresher. Further separation is done by winnowing or in an aspirated screen cleaner. Hand extraction of seed of pungent types is not pleasant for workers as they are likely to experience irritation of eyes and mucous membranes.

Where there is large-scale production of peppers for a dehydration plant, good quality seeds can be recovered by washing the separated seed from associated materials. The seed can be separated as described for tomatoes by a water trough (flume) process and then dried. Further upgrading of the clean sample can be achieved by passing it over an aspirated screen cleaner after drying. Close liaison should be maintained with the processing plant to ensure a high degree of trueness to type and no adverse high temperature effect on the seed.

Seed yield

The seed yield in any one area will usually depend on whether the cultivar is a pungent or sweet type. The pungent types are generally the higher yielding of the two. A satisfactory seed yield is from 100 to 200 kg per hectare.

A useful rule of thumb for small quantities of fruit is that 1 kg of small fruited pungent types will yield 25–100 g of seed while 1 kg of sweet or large fruited type will yield 5–50 g of seed.

1000 grain weight

The 1000 grain weight is *c.* 3.5 g for the pungent types and 5 g in the sweet types.

The main seed-borne pepper pathogens with common names of the diseases they cause

Pathogens	Common names
Alternaria spp.	Fruit rot
Cercospora capsici Heald and Wolf	Frog-eye leaf spot, fruit stem-end rot
Colletotrichum piperatum (Ell. and Ev.) Ell. and Halst	Ripe rot, anthracnose
Diaporthe phaseolorum (Cooke and Ell.) Sacc.	Fruit rot
Fusarium solani (Mart.) Sacc.	*Fusarium* wilt
Gibberella fujikuroi (Sawada) Ito. syn. *Fusarium moniliforme* Sheldon	
Phaeoramularia capsicicola (Vassil.) Deighton syns. *Cercospora capsicola* Vassil. and *C. unamunoi* Castellana	Leaf mould, leaf spot
Phytophthora capsici Leonian	*Phytophthora* blight, fruit rot
Rhizoctonia solani Kühn	*Rhizoctonia*
Sclerotinia sclerotiorum (Lib.) de Bary	*Sclerotium* rot, pink joint, stem canker
Pseudomonas solanacearum (E. F. Smith) E. F. Smith	Brown rot
Zanthomonas vesicatoria (Doidge) Dowson	Bacterial spot of fruit, stem and leaf, seedling blight
Viruses	Alfalfa mosaic virus
	Cucumber mosaic virus
	Tobacco mosaic virus

POTATO, EUROPEAN POTATO, IRISH POTATO, WHITE POTATO: *Solanum tuberosum* L.

The alternative common names of the potato distinguish it from the sweet potato, *Ipomoea batatas* (L.) Lam. to which it is not botanically related. The potato, originating from South America, is widely grown in the temperate regions of the world and is also gaining in popularity in the sub-tropics and tropics.

The commercial and home garden crops have been traditionally produced from tubers, generally referred to as 'seed potatoes', with plant breeders the main workers producing plants from true botanical seed. Reports by the International Potato Center (CIP 1979) have indicated an interest in the production of commercial potato crops from seed, especially for the tropics. The term now widely accepted for seeds of this crop is 'true seed'.

Cultivar description of potato

Season of use: suitability for specific climatic regions, daylength and market outlets

Tuber characters: shape, external colour, internal colour, characters of 'eye', including pigment.

Foliage ('haulm') relative height

Flower   colour

Resistance to specific pathogens: e.g. *Phytophthora infestans* (Mont.) de Bary.

Agronomy

There is no published work on the agronomic requirements for production of potato 'true seeds' but Accatino and Malagamba (1982) described methods of growing crops from seed by raising seedlings and transplanting, and an alternative system of direct sowing. Further work is in progress to determine suitable types for satisfactory pollination, fertilization, seed yield, high tuber yield and uniformity.

Seed extraction

The methods of extracting potato seeds from the fruits (or 'berries') is currently being investigated by several workers. A satisfactory method was described (Sadik 1982) in which the macerated berries are placed in a modified funnel similar to the apparatus used by nematologists to extract nematode cysts from soil suspensions (Fig. 12.12). The apparatus is basically a funnel

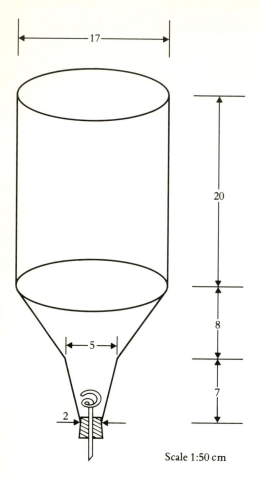

Scale 1:50 cm

Fig. 12.12   Diagram of apparatus to extract 'true potato seeds' from macerated berries (by courtesy of the International Potato Center, Peru).

with a cylinder attached and a rubber bung with coiled copper tubing fitted into the neck of the funnel. The macerated berries are placed in the top and as water passes through the coiled tubing it creates a cyclonic agitation. The scum and fruit debris overflow at the top while the seeds collect at the bottom of the apparatus and are retrieved by removing the bung.

Seed yield and 1000 grain weight

There is insufficient evidence to provide general data on seed yields but Accatino and Malagamba (1982) reported that an average flowering plant produced approximately 20 berries, each containing *c.* 200 seeds. The 1000 grain weight is *c.* 0.6 g.

The main seed-borne pathogens of *Solanum tuberosum*

The main seed-borne pathogens of *Solanum tuberosum* are viruses, the common names of the diseases they cause are:

Andean potato latent virus
Potato black ring virus
Potato spindle tuber viroid
Potato virus T
Tobacco ringspot virus
Potato virus X
Potato virus Y (potato virus X and potato virus Y together cause 'rugose mosaic').

REFERENCES

Accatino, P. and Malagamba P. (1982), *Potato Production from True Seed.* International Potato Center, Lima, Peru.

Alexander, L. J. (1962) Strains of TMV on tomato in The Netherlands and in Ohio, USA, *Meded. Landbouwhogesch. Opzoekingsstn. Staat Gent* **27**, 1020–30

Bailey, L. H. (1948) *Manual of Cultivated Plants.* Macmillan, New York.

Broadbent, L. (1976) Epidemiology and control of tomato mosaic virus, *Ann. Rev. Phytopathol.* **14**, 75–96.

Choudhury, B. (1971) Research on eggplants in India. Ford Foundation/IITA/IRAT Seminar on vegetable crops research, Ibadan, Nigeria.

CIP (1979) *Report of the Planning Conference on the Production of Potatoes from True Seed.* International Potato Center, Lima, Peru.

Cochran, H. L. (1932) Factors affecting flowering and fruit setting in the pepper, *Proc. Amer. Soc. Hort. Sci.* **29**, 434–7.

Doolittle, S. P.(1948) *Tomato Diseases.* Government Printing Office, Washington, DC.

Eguchi, T., Matsomura, T. and Ashizawa, M. (1958) The effect of nutrition on flower formation in vegetable crops, *J. Amer. Soc. Hort. Sci.* **72**, 343–52.

George, R. A. T., Stephens, R. J. and Varis, S. (1980) The effect of mineral nutrients on the yield and quality of seeds in tomato. In *Seed Production*, P. D. Hebblethwaite (ed.), pp. 561–7. Butterworths, London and Boston.

Grubben, G. J. H. (1977) *Tropical Vegetables and Their Genetic Resources*, H. D. Tindall and J. T. Williams (eds). International Board for Plant Genetic Resources, FAO, Rome.

Heiser, C.B. and Smith, P. G. (1953) The cultivated capsicum peppers, *Econ. Bot.* **7**, 214–27.

Jenkins, J. A. (1948) The origin of the cultivated tomato, *Econ. Bot.* **2**, 379–92.

Kingham, H. G. (ed.) (1973) *The Tomato Growers Handbook.* Grower Books, London.

Lesley, J. W. (1924) Cross pollination of tomatoes, *J. Heredity* **15** 233–5.

Luckwill, L. C. (1943) The genus *Lycopersicon* an historical, biological and taxonomic survey of the wild and cultivated tomatoes. *Aberdeen University Studies*, No. 120. The University Press, Aberdeen.

Murthy, N. S. R. and Murthy, B. S. (1962) Natural cross pollination in chilli, *Andhra Agric. J.* **9**(3), 161–5.

Nitzany, F. E. (1960) Transmission of tobacco mosaic virus through tomato seed and virus inactivation by methods of seed extraction and seed treatments, *Ktavim Rec. Agric. Res. Sta.* **10**, 63–7.

Purseglove, J. W. (1974) *Tropical Crops, Dicotyledons.* Longman, London.

Richardson, R. W. and Alvarez, E. L. (1957) Pollination relationships among vegetable crops in Mexico I. Natural cross pollination in cultivated tomatoes, *Proc. Amer. Soc. Hort. Sci.* **69**, 336–71.

Rick, C. M. (1978) The Tomato, *Scientific American*, **239(2)**, 67–76.

Rylski, I. (1972) Effect of early environment on flowering in pepper (*Capsicum annuum* L.), *J. Amer. Soc. Hort. Sci.* **97**(5),648–51.

Rylski, I. (1973) Effect of night temperature on shape and size of sweet pepper (*Capsicum annuum* L.), *J. Amer. Soc. Hort. Sci.* **98**(2), 149–52

Sadik, S. (1982) A method for seed extraction in *True Potato Seed (TPS) Letter,* **2**(3).

Seth, J. N. and Dhaudar, D.G. (1970). Effects of fertilizers and spacings on the seed yield and quality in brinjal (*Solanum melongena* L.) variety Pusa Purple Long, *Progressive Horticulture* **1**(4), 45–50.

FURTHER READING

Abdel-Wahar, A. E. and Miller, J. C. (1963) Re-evaluation of some techniques and their effect on stimulating flowering in four Irish potato varieties in Louisiana, *Amer. Potato. J.* **40**, 53–7.

Barker, W. G. and Johnston, G. R. (1980). The longevity of seeds of the common potato, *Solanum tuberosum, Amer. Potato J.* **57**, 601–7.

Henderson, M. T. and LeClerg, E. L. (1943) Studies of some factors affecting fruit setting in *Solanum tuberosum* in the field in Louisiana, *J. Agric. Res.* **66**, 67–76.

Kuo, C. G., Chen B. W., Chou, M. H., Tsai, C. L. and Tsay, T. S. (1979) Tomato fruit-set at high temperatures, *Proc. First International Symposium Tropical tomatoes*, pp. 94–108. AVRDC, Taiwan.

Pushkarnath, I. M. and Chauhan, H. S. (1964) Effect of gibberellic acid and temperature treatments on flowering and seed setting in some potato varieties, *Indian Potato J.* **6**, 90–5.

Rick, C.M. (1978) The tomato, *Scientific American* **239**(2), 67–76.

Thijn, G. A. (1954) Observations on flower induction with potatoes, *Euphytica* **3**, 28–34.

Villareal, R. L. (1980) *Tomatoes in the Tropics.* Westview Press, Boulder, Colorado

13 UMBELLIFERAE

There are many genera in this family which are grown as vegetables for cooking, salads, condiments or flavourings. The main vegetables whose seed production is discussed below are:

Daucus carota L. subsp. *sativus* (Hoffm.) Thell.	carrot
Pastinaca sativa L.	parsnip
Petroselinum crispum (Mill.) Nym. ex A. W. Hill	parsley
Apium graveolens L. var. *dulce* (Mill.) DC. syn. *Apium dulce* Mill.	celery
Apium graveolens L. var. *rapaceum* (Mill.) DC.	celeriac, turnip rooted celery

These five vegetable crops are extremely important in temperate regions of the world. Celery and carrot are especially popular in North America and Europe but they are also important locally in other areas of the world including some sub-tropical regions.

Other umbelliferous crops are either of local importance only as vegetables or salads, or are used as flavourings, garnishes or condiments. The following crops are usually grown from seed:

Anthemum graveolens L.	dill
Anthemum sowa Kurz.	Indian dill
Anthriscus sylvestris Hoffm.	chervil
Carum carvi L.	caraway
Coriandrum sativum L.	coriander
Cuminum cyminum L.	cumin
Foeniculum vulgare Mill.	fennel
Foeniculum vulgare Mill. var. *dulce* (Mill.) Thell.	Florence fennel
Pimpinella anisum L.	anise

Arrocache (*Arracacia xanthorrhiza* Bancr.) is a starchy perennial root crop which is important in parts of South America, but is usually propagated vegetatively.

CARROT: *Daucus carota* L. subsp. *sativus* (Hoffm.) Thell.

Origins and types

The modern cultivated carrot has been derived from the wild carrot *Daucus carota* L. found in Europe, Asia and Africa.

There appear to be two main sources of material from which early selections were made; these are the anthocyanin carrots from Asia (especially centred in Afghanistan where purple-rooted carrots are still cultivated) and the carotene carrots developed originally in Europe. Banga (1963) describes the descent and development of the main types of what he calls the 'Western carotene carrot'.

Carrot cultivars are normally classified according to root shape and size. Typical root shapes of cultivars grown in Northern Europe are shown in Fig. 13.1. Further divisions can be made according to their season of use and root colour. Banga (1964) has described the types of orange rooted carrots and their specific characters. In the UK the National Institute of Agricultural Botany (NIAB) publishes a list of maincrop carrots annually (Anon. 1983).

Fig. 13.1   Types of carrot roots, left to right: short horn, Nantes, Amsterdam forcing, Chantenay, Autumn King, and St Valery.

Cultivar description of carrot

Season   early (suitability for protected cropping or production under polythene) or maincrop

Type   according to basic root type

Use   bunched or bare roots for fresh market, washed and topped for pre-packing
            processing: canning, freezing or dehydration (high dry matter

content for dehydration)
storage: lifted or left in ground

Foliage   amount of foliage, degree of fineness of petiole bases, colour of
petiole bases

Root   length and relative size
shape and overall outline and shape of shoulder, shape of root
tip
surface, degree of smoothness or roughness
colour, shoulder
internal root colour (core colour)

Flowering

Flower initiation

The carrot cultivars which originated in Asia tend to be annuals when grown
in long days and do not have a vernalization requirement.

The cultivars which have been developed in the temperate regions of Europe and North America are biennials which produce a swollen tap root at
the end of the first season with a rosette of leaves. The plant generally requires a cold period before it will bolt. The response to cold is not necessarily
related to root size. Even in the types which require a cold period to flower,
some plants occur which have the annual habit; these plants, which bolt in
their first year, also have poor roots and are rejected during the course of
cultivar maintenance.

Pollination

Individual carrot flowers are normally protandrous and much cross-pollination occurs between plants in a seed crop. However, because of the extended
flowering period resulting from several successive umbels per plant and the
succession of flowers on individual umbels, the possibility of self-pollination
always remains.

Pollinating insects

Bohart and Nye (1960) observed the occurrence of pollinating insects on
flowering carrots in Utah and noted that while honeybees were efficient pollinators they were frequently scarce on carrot crops because other crop species
were flowering in the vicinity at the same time. They also observed that several insect genera in Hymenoptera, Diptera and Coleoptera were extremely
important pollinators of carrots, in the absence of bees.

Hawthorn *et al.* (1960) found that when there was an adequate supply of
pollinating insects both seed yield and seed quality were high; where natural
insect pollinator populations were low there was an advantage in increasing
the number of honeybee colonies by placing groups of hives adjacent to the
carrot fields.

Isolation

Because of the high possibility of cross-pollination, isolation distances for commercial seed crops should be a minimum of 800 m. For basic seed the distance should be greater, about 1600 m.

In areas which specialize in carrot seed production the different cultivars within the same type can be zoned; this minimizes pollen contamination between the different types.

Cultivated carrots cross-pollinate very readily with the wild carrot and this must be taken into account when choosing sites for seed production. Contamination of seed crops by wild carrot pollen is a major reason for genetical deterioration of seed stocks in some areas of the world.

Order of flowering

The individual carrot flowers, in common with most other species in Umbelliferae, are borne on terminal branches in compound umbels. There is a distinct order of flowering which relates to umbel position. The first umbel to flower is the primary (sometimes referred to as the 'king' umbel) which is terminal to the main stalk. Branches from the main stalk form secondary umbels, and subsequent branches from these form tertiary umbels. Quaternary branches and umbels may also be formed. The relative positions of primary, secondary and tertiary umbels are shown in Fig. 13.2.

There is, therefore, a wide range of flowering time and seed maturity on each plant. The relative position of seeds, i.e. the umbel on which they are formed, has been the subject of research as there are correlations between umbel order and seed quality which are discussed later. Other sources of variation in time of flowering between plants have been shown to be related to size of steckling.

The modern methods of carrot production for the fresh market and processing require high quality carrot seeds with the minimum of variation between individual plants derived from the same seed-lot. The main sources of variation in carrots and the implications for changes in seed production techniques have been reviewed by Gray and Steckel (1980).

Plant population and seed yield

Carrot seed yield increases with plant population; in the 1950s Hawthorn (1951) and Hawthorn and Pollard (1954) demonstrated that average seed yields per annum increased from 580 kg per hectare to 1039 kg per hectare as the number of stecklings planted increased from 12 000 to 140 000 per hectare. More recent work by Gray (1981) extended the concept as suggested by earlier workers that an even greater seed yield could be obtained by further increasing the plant population per unit area. Gray investigated the possibility of increased plant densities in root to seed method from 100 000 to 800 000 stecklings per hectare, and from 110 000 to 2 560 000 plants per hectare by the seed to seed. Seed yields increased from 1250 to 2000 kg per hectare for

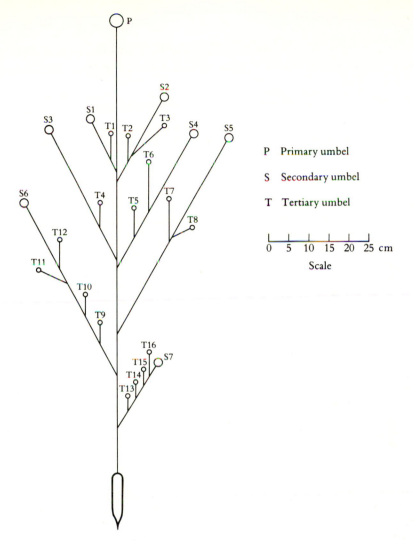

P Primary umbel

S Secondary umbel

T Tertiary umbel

Fig. 13.2   Diagram showing umbel order and positions in carrot (from Sandin 1980).

the root to seed method and from 700 to 2400 kg per hectare for the seed to seed (Fig. 13.3).

The number of umbels per plant is decreased with increasing plant density. Gray found that at the highest densities the umbels matured more or less simultaneously, but at the lowest densities used in his experiments the seed in the primary umbels started to ripen approximately two weeks before seed in the secondary umbels. Jacobsohn and Globerson (1980) investigated four plant densities and found that the highest seed yield and the highest proportion of primary umbels were obtained from their highest mother plant population (Table 13.1).

In practice increasing plant densities in mechanized systems depends on the

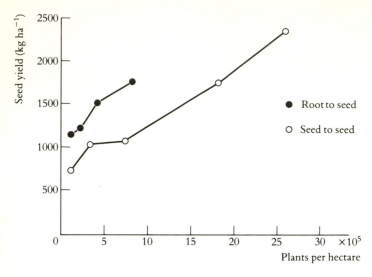

Fig. 13.3　Response of carrot seed yield to plant density (from Gray 1981).

Table 13.1　Effect of plant spacing on seed yield of carrot plants (from Jacobsohn and Globerson 1980)

Spacing (cm)	Total yield (g)		Yield of germinated seeds★		Seeds from primary umbel (%)
	(per plant)	(per m²)	(per plant)	(per m²)	
80 × 40	5.7[ab†]	17.8[c]	2.3[b]	7.2[c]	25[c]
80 × 7	8.8[a]	157.7[b]	3.2[a]	56.4[b]	37[b]
40 × 7	3.5[b]	125.3[b]	1.6[c]	56.4[b]	50[a]
20 × 7	3.4[b]	240.8[a]	1.4[c]	98.4[a]	51[a]

★ Calculated on the basis of seed germination in the laboratory.
† Means within columns followed by different letters, are statistically different at $p = 0.5$.

ability of transplanting machines to plant stecklings at densities greater than their present capabilities of 100 000 plants per hectare.

The main advantage of the higher plant densities is probably the shortening of the overall flowering period and increased evenness in umbel ripening. A uniform seed crop facilitates the use and timing of pre-harvest desiccant sprays to accelerate drying, and the use of PVA adhesives which reduce pre-harvest seed drop from the umbels. Other advantages suggested by Gray include timing of cutting the crop to obtain maximum seed yield. The high plant densities tend to produce a concentration of umbels in the upper part of individual plant stalks; thus if the cutter bar can be raised during cutting, the subsequent windrows rest on longer stubble, allowing increased air circulation under them prior to threshing.

The high plant densities also improve seed quality because there is less branching and therefore a higher proportion of primary to secondary umbels than at lower plant densities. This is illustrated in Fig. 13.4.

A further advantage is that overall seed quality is improved by a shorter time

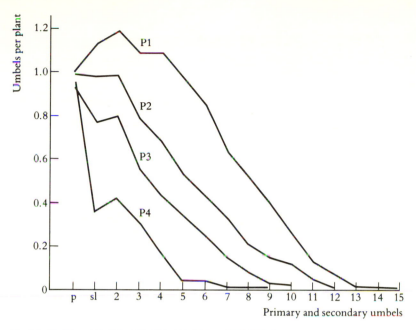

P1, P2, P3, P4 = 100 000, 200 000, 400 000 and 800 000 plants per hectare respectively.

Fig. 13.4   Effect of carrot plant density on branching in the seed crop (from Gray 1981).

span of seed maturity on the first and last umbels of each plant, shortening cutting time prior to placing in windrows.

Influence of seed position on mother plant on seed quality and seedling performance

Germination and seedling emergence evaluations were made by Thomas *et al.* (1978) with carrot seed harvested separately from primary and secondary umbels when either 'immature' (i.e. forty-seven days after anthesis) or 'mature' (i.e. sixty-eight days after anthesis). Germination was approximately 20 per cent lower in the 'immature' than in the 'mature' seeds at temperatures below 12 °C and greater than 25 °C. 'Immature' seeds from secondary umbels took approximately six days longer to germinate at 5 °C than those from primary umbels. In addition, the weights of seedlings from primary umbel seeds were greater than those from secondary umbel ones when weighed at about twenty days after 50 per cent seedling emergence.

This interesting result was substantiated by Gray (1979) who showed clearly the relationship between umbel position and seed quality (Fig. 13.5). This work has resulted in increases in plant density in commercial seed production. Because the primary umbel seed is ready for harvest some time in advance of umbels of a lower order, especially at low plant densities, hand-harvesting operations have been used in some areas of the world (Fig. 13.6). In mechanized production systems changes are taking place in planting systems to increase plant density.

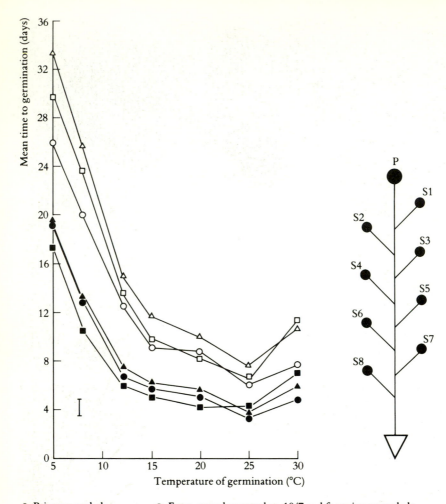

○ Primary umbels   ● From crops harvested on 10/7 and for primary umbels
□ Secondary 1 umbels   ■ Secondary 1 umbels
△ Secondary 8 umbels   ▲ Secondary 8 umbels from crops harvested on 7/8;
                             I = L.S.D. ($p = 0.05$).

Fig. 13.5   Relationship between mean time to germination and temperature for different umbel positions (from Gray 1979).

Nutrition

Little experimental work has been done on the nutrition of carrots for seed production. Most of the commercial seed producers rely on either personal observations or recommendations based on the production of marketable root crops when preparing soils for the first season of steckling production. These are usually nitrogen to phosphorus and potassium ratios of approximately 1 : 2 : 2 N : P : K, the phosphorus application being lowered for soils with a high phosphorus status.

Fig. 13.6   Hand harvesting primary umbels of carrot.

Experimental work by Austin and Longden (1966) indicated that improved seed quality resulted from increased soil fertility achieved by inorganic fertilizers and organic manures applied prior to the year of flowering.

Cooke (1975) suggested that there is a clear case for supplying sodium to carrots for root production and reported that in the UK 380 kg per hectare of sodium chloride increased yields to the same extent as 190 kg per hectare of potassium chloride. But as far as the author is aware there has been no experimental work including sodium fertilizers in nutrient experiments for carrot seed production.

Top dressings of up to 200 kg per hectare of nitrogen are given in the spring of the second year by some carrot seed producers. Nitrogen applied in this way will counteract losses from leaching especially in areas of high winter rainfall.

Agronomy

There are two systems of carrot seed production, 'seed to seed' and 'root to seed'.

'Seed to seed' production

In this method the stock seed is sown in the late summer (July or early August in the northern hemisphere) and grown *in situ* through the following winter. The plants flower the following spring and seed is harvested in the late summer of the year following sowing.

Row spacings of 50–90 cm are used, with a sowing rate of 2–3 kg per hectare. The roots cannot be inspected or rogued and in practice virtually no roguing is done for foliage characters either; the system relies therefore on very good quality stock seed and satisfactory isolation.

'Root to seed' production

This system is similar to the 'seed to seed' system as far as timings are concerned but the plants ('stecklings') are raised in beds and transplanted in the spring. Depending on local customs and winter conditions the stecklings are either left *in situ* during the winter or lifted in the late autumn and stored until re-planting in the spring.

The ratio of steckling bed to final area transplanted is from 1 : 5 to 1 : 10, depending on the degree of roguing, if any, prior to storage or transplanting into the final quarters and also the degree of uniformity of transplant used.

The transplanted steckling rows are 75–90 cm apart with 20–30 cm between plants. The sowing rate is 6–8 kg per hectare for the root bed. Raising stecklings which are later transplanted from their beds offers the opportunity of roguing plants with undesirable root or foliage characters when lifting and planting, but the system is very labour intensive and its use is declining. One advantage is that the area of land used in the first year is significantly less than the area subsequently planted for seed production.

A variation of this method involves the production of mature roots as for fresh markets. The roots are lifted in the autumn, rogued on root characters and stored prior to planting in the late winter. Some growers mow off the carrot tops before lifting, but care must be taken not to damage the crowns of the roots. About 5–8 cm of the leaf tissue is left on the roots. In some areas the roots are re-planted without storage. In this case it may be necessary to protect the transplanted roots from frost by applying a layer of litter or straw to the surface of the beds.

It is important that any diseased or mechanically damaged roots are discarded during the roguing of the lifted roots. Roots should be relatively free from soil, but not washed, before storage. The method of storage depends to some extent on local traditions. In Scandinavia straw- and soil-covered pits or 'clamps' are used but in other areas barns or purpose-built stores are used. Fine soil, moist peat or sand is placed between layers of roots.

The ideal storage temperature is 1 °C with humidity at about 90–95 per cent. Controlled ventilation stores use outside air for cooling when the ambient temperature is above freezing.

Planting out
Roots (or stecklings) are planted out from the store in the early spring when soil conditions are satisfactory. The roots must not be allowed to dry out at this stage as this adversely affects establishment and subsequent seed yield.

Large-scale seed producers use planting machines whereas on a small scale dibbers, trowels or spades are used. Whatever planting system is adopted, the crown of the root must be at or just below the final surface of the firmed soil in which it is planted. The post-planting time is critical for quick plant establishment, and if conditions are dry, irrigation should be applied. It is useful in some areas to earth the rows up with about 10 cm of soil when the plants are re-established.

Harvesting seed

In small-scale commercial production where hand labour is plentiful the umbels are cut by hand as they ripen. This system is also used for small plots of high value stock seed.

On a larger scale, where mechanized systems are used, it is necessary to decide the best cutting time for maximum seed yield. There is a tendency for some carrot seed to be lost by shattering or dropping from primary umbels if cutting is delayed to allow maturity of seed on secondary and tertiary umbels. In some areas of the world the plants are sprayed with polyvinyl acetate to prevent loss from the primary umbels. Harvesting early in the day when the dew is still on the seed heads will also reduce the amount of seed which is lost from the umbels.

Generally, the crop is cut when the earliest maturing seed on the primary umbel is mature and starting to drop. Carrot seed is brown when ripe and at this stage the umbel is brittle. The cut crop continues to dry in windrows.

When the seed crop has ripened it can be separated from the remainder of the plant debris by a combine harvester. As with the initial cutting, the combining can be done early in the day to take advantage of dew reducing loss from dropping. The plant material must not be too dry or in addition to losing seed, the plant debris will shatter and increase the subsequent seed cleaning operation. Where the scale of operation is inappropriate for combine harvesters the dry material is put through a thresher.

Carrot (like dill and caraway) has spines or 'beards' on the seed. These must be removed by a debearder before further cleaning operations. Debearding improves the seed flow and reduces the volume of the seed-lot. Figure 13.7 shows carrot seed before and after debearding. Further cleaning is achieved by aspirated screens and indent cylinders.

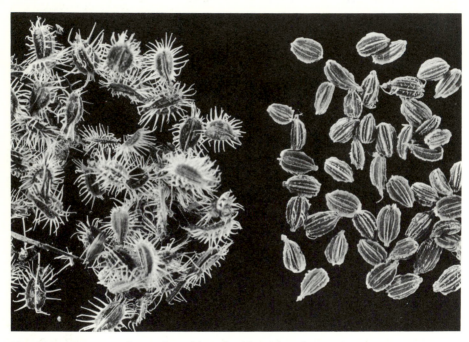

Fig. 13.7   Carrot seed samples with and without beards.

Seed yield and 1000 grain weight

The yield of carrot seed at different plant densities is the subject of current investigations in several research centres of the world. At present expected yield of open-pollinated cultivars in the temperate regions is about 600 kg per hectare with highest yields achieved reaching 1000 kg per hectare. The 1000 grain weight is *c.* 0.8 g.

Yields in the tropical regions are usually considerably lower despite using higher altitudes to achieve satisfactory vernalization, and figures of about 300 kg per hectare should be expected for the European types. The Asiatic types produce only about 250 kg per hectare when seeded in the tropics.

Production of F₁ hybrid carrot seed

Some seed companies have developed inbred lines and use a suitable male sterile line as the female parent for the production of F_1 hybrid seed. Generally the ratio of pollinator rows to females for seed production is 2 : 4 or 2 : 6.

Seed yields of F_1 hybrid carrot seeds are relatively low, about 500 kg per hectare, and after taking into account the exclusion of the seed from pollinator rows the lower yield is frequently attributed to less insect activity because of smaller petals on the male sterile flowers of the seed-producing lines. F_1 hybrid carrot seed producers frequently use colonies of bees to supplement the natural level of pollinating insect activity.

Carrot roguing stages and main characters to be observed

'Seed to seed' production

Very little if any roguing can be done when the crop is grown-on without lifting. But plants bolting early and those with untypical foliage characters should be removed.

If the crop is lifted and re-planted it is rogued as described below for 'root to seed' but very little confirmation of root type can be done.

'Root to seed' production

During the first year's growing season:

1. Remove plants displaying atypical foliage. Remove plants bolting in the first year.
2. After the roots have been lifted inspect for trueness to type, according to root shape, colour and size. Discard roots showing poor colour, incorrect colour, coloured shoulders (purple, green), split, fanged, rough surface.

Figure 13.8 shows the range of root defects which should be rogued out when inspecting lifted root crop.

Improvement of basic stock

Grow the crop on the 'root to seed' system. Select roots to a very high standard of trueness to type and freedom from defects before storage.

Inspect the roots again after storage, especially to remove roots showing storage diseases.

Where necessary for stocks requiring high core colour, check selected roots by either cutting a small portion off the tip of the root, or take a small hole with a cork borer immediately before planting. Treat the cut surface with a suitable fungicide powder and plant immediately.

For stock seed all plants can be bulk harvested; progeny testing can be done before bulking.

Fig. 13.8 Range of carrot root defects.

The main seed-borne carrot pathogens with common names of the diseases they cause

Pathogens	Common names
Alternaria dauci (Kühn) Groves & Skolko	Carrot leaf blight
Alternaria radicina Meier, Drechsl. and Eddy syn. *Stemphylium radicinum* (Meier, Drechsl. and Eddy) Neergaard	Black root rot, seedling blight
Cercospora carotae (Pass.) Kazn and Siem.	*Cercospora* blight of carrot, leaf spot
Gibberella avenacea Cook syn. *Fusarium avenaceum* (Fr.) Sacc.	Brown root rot
Phoma rostrupii Sacc.	*Phoma* root rot
Xanthomonas carotae (Kendrick) Dowson	Bacterial blight, root scab
Viruses	Carrot motley dwarf (a complex of three viruses, including carrot red leaf)
	Carrot red leaf virus

PARSNIP: *Pastinaca sativa* L.

Origins and types

Parsnips are essentially a European vegetable but are grown as a relatively minor crop in North America and have also been introduced to other temperate parts of the world.

Traditionally the root is used for culinary purposes in the autumn and winter at the end of the first season from seed. However, with modern pre-packaged marketing there is a tendency for roots to be harvested over a longer period, commencing in the summer with young, small roots grown at close spacings.

There is a relatively small range of root types represented in the parsnip cultivars available. The basic shapes are wedge, bayonet and bulbous (see Fig. 13.9). In recent years plant breeders have put emphasis on producing material which is resistant to canker.

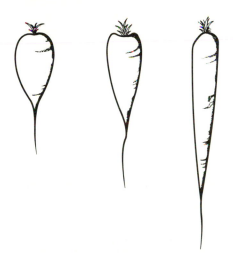

Fig. 13.9   Parsnip root types; left to right, bulbous, wedge and bayonet.

The parsnip plant is a biennial which produces a swollen tap root with a rosette of leaves. The root is fully developed by the end of its first season when the leaves die down. It flowers in the second year following natural vernalization of the roots during the winter. No research work has been reported on the low temperature requirement, but in Northern Europe the dormant roots of parsnip are always vernalized by the end of the winter and new leaves are produced from the crown of the root by early spring.

Cultivar descriptions of parsnip

The cultivar descriptions of parsnip are normally based on root characters.
Use   season: processing (canning), fresh market, pre-packaging
Root   shape: bulbous, wedge, bayonet or intermediate between two

length
size
crown: shallow, hollow (deep)
shoulder: square or round
surface: texture: smooth, rough or intermediate
surface colour: white or cream
resistance to canker
other root characters: relative resistance to bruising in transit, in-
    cidence of fanging

Agronomy

The roots of parsnip are not sensitive to frost and it is therefore possible to produce a seed crop by the 'seed to seed' method without the need to lift, store and re-plant the roots. However, for basic seed and other high genetic quality stock seed production it is vital that the 'root to seed' method is used so that the choice of plants for seeding is based on root characters.

The parsnip root weight and size are usually directly related to the length of time it is growing in its first year. Therefore, in order to make selections of roots in the autumn which are typical of the cultivar, it is necessary to sow early in the spring.

Germination and seedling emergence can take up to three weeks and this is another factor contributing to the need for relatively early sowing for the 'root to seed' method.

Soil type

The best quality roots are produced on deep peat, silt or sandy loam soils. The presence of stones will cause forking or 'fanging' of the roots (see Fig. 13.10) and shallow or stony soils should be avoided if roots are to be lifted for selection.

Nutrition and pH

This crop requires a soil with pH around 6.5; it will not succeed when the pH is less than 6.0.

A fertilizer with a ratio of 1 : 2 : 2 N : P : K is applied during the final preparation of the seed bed, according to the nutrient status of the soil.

Many parsnip seed producers apply a top dressing of a nitrogenous ferti-lizer in the spring following vernalization once the plants are seen to be pro-ducing leaves. This can be applied as ammonium sulphate at a rate of 200 kg per hectare. The top dressing is especially useful in areas where there is a high leaching rate during the winter or early spring.

Plant production

'Seed to seed' method

Seed is sown in rows 1 m apart in early summer, usually between mid-May and mid-June. At this row spacing the sowing rate is 4 kg per hectare. The

Fig. 13.10   Fanging of parsnip roots.

later sowings may well run into moisture deficit problems for satisfactory germination. In North America irrigation is usually necessary to ensure satisfactory seedling emergence. Because of the relatively slow germination the seed should either be sown in a 'stale seed bed' or a suitable pre-emergence residual herbicide such as Linuron applied.

The seedlings are thinned to a distance of 60 cm if the plants are to flower *in situ*; if they are to be lifted for root inspection then half this distance in the row is satisfactory but the plants are re-planted at the greater distance after lifting for inspection.

'Root to seed' method

The same system is followed as for the 'seed to seed' method except that if stock seed is being produced the sowing date should be in accord with the sowing date for the fresh market crop. This will ensure that the roots have reached their typical size and shape as well as showing other characters when they are lifted for selection in the autumn.

The sowing rate is similar to the 'seed to seed' rate, but growers who use a closer row spacing use an increased sowing rate. This will depend on the machinery used for lifting the roots.

The roots are lifted in the late autumn or early in the new year according to local custom based on whether or not the ground is regularly frozen in mid-winter.

The method used for lifting roots depends on the scale of operation. In large-scale production they are ploughed out, but on a smaller scale, especially for production of basic seed, the roots are forked out. The roots are examined for trueness to type as soon after lifting as possible and must not be allowed to dry out.

After selection, the roots to be retained for seed production the following season are re-planted, usually in rows 1 m apart with 30–60 cm between plants within the rows. If the ground is frozen, small quantities of roots selected for basic or breeders' seed can be stored in moist soil, sand or peat until planted.

No spacing experiments have been reported with parsnip seed production, but as with carrots there is a trend to higher density crops in order to increase the ratio of primary umbels to secondary and tertiary umbels.

Pollination

Parsnip is predominantly cross-pollinated. The flowers are protandrous which reduces the chance of selfing of individual plants but because of the umbel succession some selfing occurs. There is a tendency for the last flowers on the ultimate umbels not to be fertilized. (This must be taken into account when choosing the cutting time.)

Pollination is by a wide range of insect species including Diptera and Lepidoptera but bees are the main pollinators. Seed producers frequently supplement natural insect populations with colonies of hive bees. The honey produced by bees working parsnip flowers is reputed to be dark in colour and of very good quality.

Small enclosed lots of flowering parsnips in cages or structures are generally pollinated by blowflies put into the enclosure at regular intervals according to the life of the insects.

Isolation distances should be a minimum of 500 m between parsnip seed crops, including commercial root crops which may contain a small percentage of bolters in their first year.

Harvesting

Parsnip plants grown on for seed production are very tall and the mature plant with umbels can reach more than 2 m in height. Crops produced on the 'seed to seed' system are generally taller than 'root to seed' crops. Plants are also generally taller at greater than lower plant densities but, as with carrots, closer spacing will increase the proportion of seed from primary umbels.

The plants are cut when the majority of the seeds in the primary umbel have ripened. This can be detected by their light brown colour and by the splitting of the schizocarp to display two separate mericarps which are the 'seed'. Parsnip seed 'shatters' readily, and timing the cutting of the crop can be critical especially if there is the likelihood of inclement weather. As with carrots some producers have used PVA glues to reduce loss by shattering.

Small areas of parsnip seed can be cut by hand but larger areas are cut with a mower and left in windrows to dry. In some areas of North America, especially Oregon, the swather is taken over the field twice. The first cut is at about 75 cm high which leaves the cut material perched on a high stubble. This is followed by a second cut of the stubble under the swath which brings all the material to ground level.

The seed can be threshed by a combine or stationary thresher, but as the dead or dry plant material (especially umbels) fractures easily and will add to the plant debris which is later difficult to separate from the seed, the material is threshed when the crop is still damp from dew.

Some workers are susceptible to the juices of parsnip plants which can cause blisters on the exposed skin, so it is advisable to wear protective overalls, gloves and goggles when handling the crop and to seek early medical advice for workers who show signs of an allergy.

Final cleaning of the seed can be achieved by passing the material through aspirated screen separators.

Seed yield and 1000 grain weight

The average seed yield is about 1000 kg per hectare, but some parsnip seed producers in Oregon, USA, can achieve double this. Seed weight: the 1000 grain weight of parsnip seed is approximately 1 g.

Roguing of parsnip

The roguing or selection of parsnip is normally only done when the roots have been lifted. Thus the 'seed to seed' crops are not inspected. Plants showing atypical characters should be discarded. Plants which bolt in their first year and umbelliferous weeds are removed during roguing.

Parsnip root characters to be observed

Shape   relative length/width, shape of shoulders

Crown   degree of depression

Quality   incidence of fanging (or forked roots). This can be affected by soil type, but fanged roots should be rejected. Freedom from splitting

Colour   white, cream

Surface texture   smooth

Resistance to diseases responsible for canker, i.e. *Itersonilia pastinacea*, *Phoma* spp. and *Centrospora acerina*.

Bolting plants: uniformity of height and colour of individual plants.

The main seed-borne parsnip pathogens with common names of the diseases they cause

Pathogens	Common names
Alternaria dauci (Kühn) Groves and Skolko syn. *A. porri* (Ell.) Cifferri f.sp. *dauci* (Kühn) Neergaard	
Alternaria radicina Meier, Drechsl. and Eddy syn. *Stemphylium radicinum* Meier, Drechsl. and Eddy	Black mould
Erysiphe heraclei DC. ex St Amans	Powdery mildew
Itersonilia pastinacea Channon	Canker
Virus	Strawberry latent ringspot virus

PARSLEY: *Petroselinum crispum* (Mill.) Nym. ex A. W. Hill

Origins and types

This crop is grown for its foliage which is used fresh or dried as a herb, flavouring, garnish or fresh as a green salad.

The wild species is a native of Europe but cultivated forms are now spread throughout the world in temperate and tropical regions. There has been interest from time to time in the fresh leaves as a source of vitamin C but most of the current production is for dehydration.

Parsley is a biennial but when grown as a commercial foliage crop the leaves are utilized in the first year and the crop is treated as an annual. The seed crop is vernalized during the winter at the end of the first season.

There are two main types of parsley, based on foliage characters which are curled or plain. There are many selections of the two types, based on leaf colour (intensity of chlorophyll), degree of curliness and petiole length (Fig. 13.11).

Parsley cultivar descriptions

These are based on foliage characters (except for Hamburg parsley) and include the intensity of green colour, petiole length and leaf crenulation.

Pollination

Parsley is predominantly cross-pollinated by a wide range of insects including honeybees and some genera of Diptera. Isolation should be at least 500 m between cultivars of the same leaf type, but increased to 1000 m between curled and smooth leaved cultivars.

Fig. 13.11 A plain leaved off-type in a curled leaf parsley crop.

Agronomy

Parsley seed production requires similar soil types and nutrients as carrot.

'Seed to seed' production

The seed is sown at a rate of 3 kg per hectare in rows of 56 cm apart during July or August. Parsley seed takes up to 30 days to germinate under field conditions.

'Root to seed' production

Seed is sown in May or June. The higher sowing rate of up to 6 kg per hectare is used for this method. Seed is sown in rows at about half the distance apart as the 'seed to seed' system, the exact distance depending upon the presence of irrigation furrows and the need for mechanical cultivations when residual herbicides are not used. An area of plants produced from the above sowing rate will provide sufficient transplants for up to thirty times the area when transplanted at 90 cm. There is, however, a trend towards closer planting distances in order to reduce the time span of seed maturing from successive umbels on individual plants.

Only the 'root to seed' system should be used for production of basic and other stock seed; this ensures that any early bolting plants can be rogued.

Harvesting

The ripe seed of parsley shatters easily and it is therefore very important that the crop is cut at a stage to minimize loss, usually just before the primary umbels shatter.

The crop which is between 1 and 2 m tall by the time that flowering finishes is cut into windrows. Large-scale producers use a combine for separating the seed from the straw, but on a small scale the seed is separated by using stationary or mobile threshers.

Further separation and cleaning is later achieved by passing the seed through an aspirated screen cleaner.

Hamburg parsley

This is a selection of smooth leaved parsley with a swollen tap root. Both the root and foliage are used for culinary purposes. The root, which is similar in size to those of the shallower parsnip cultivars, has a very strong tendency to fanging and it is therefore important to produce seed by the 'root to seed' system so that individual plants are selected on root quality.

Seed yield and 1000 grain weight

The average parsley seed yield is approximately 800 kg per hectare, but it is probably possible to achieve higher yields than this at higher plant densities with a larger proportion of seed ripening simultaneously. Seed weight: the 1000 grain weight of parsley seed is 2 g.

Current trends

In the USA some large-scale parsley seed producers take a second seed crop in the third year. This saves the cost of sowing and young plant management but can result in a high weed seed content if herbicides are not used.

An interesting innovation in the USA, especially in Oregon, is parsley for seed production grown with an intercrop of dill (*Anthemum graveolens* L.). Dill is an annual and the two species are sown together in the spring by a modified seed drill with two seed boxes. Seeds of the two species are sown in the same drills. The advantage of this intercropping is that seed of the two species is harvested in separate years. The higher leaf canopy of dill is relatively thin and sufficient light reaches the parsley. The dill seed is cut and removed with its trash in the autumn of the first year but the parsley rosettes are relatively undamaged. The parsley plants flower in the second year and their seed is then harvested. This is a rare example of intercropping in mechanized vegetable seed production. The two species are resistant to the same herbicides, and producers maintain that energy, space and herbicides are saved.

Some large-scale parsley seed producers in the UK undersow the parsley with barley in the first year. This also provides an income during the year that the parsley is in the vegetative phase, but relies on high quality stock seed rather than roguing for maintaining genetical quality.

Parsley roguing stages

1. Young plants in rosette stage before first autumn. Plants with annual habit (bolters in first year) must be removed.
   Leaf characters
   Colour
   Petiole length
   Crenulated or smooth
   Degree of crenulation and division of leaf
2. Lifted plants, Hamburg parsley only, roots relatively free from fanging.
3. Second year. Leaf characters as above.

The main seed-borne parsley pathogens with common names of the diseases they cause

Pathogens	Common names
Alternaria dauci (Kühn) Groves and Skolko syn. *A. porri* (Ell.) Ciferri f. sp. *dauci* (Kühn) Neergaard	
Alternaria radicina Meier, Drechsl. and Eddy syn. *Stemphylium radicinum* (Meier, Drechsl. and Eddy) Neergaard	
Erysiphe heraclei DC. ex St. Amans syns. *E. umbelliferarum* de Bary. and *E. polygoni* DC. em. Salm. pro parte	Powdery mildew
Gibberella avenacea Cook syn. *Fusarium avenaceum* (Fr.) Sacc.	Brown root rot
Phoma anethi (Pers. ex Fr.) Sacc. syns. *Marssonina kirchneri* Hegyi and *Passalora kirchneri* (Hegyi) Petr.	Leaf and stem spot
Rhizoctonia solani Kühn	Root and basal stem rot
Septoria petroselini Desm.	Leaf spot

CELERY: *Apium graveolens* L. *dulce* (Mill.) DC.

Origins and types

The modern cultivars of celery, which are popular salad and cooking crops in Europe and North America, have been developed from a wild marsh plant which is widely distributed in Europe and Asia.

There are two basic groups of celery cultivars: trench and self-blanching. The trench types are usually planted in shallow trenches and earthed-up ('blanched') during the latter part of their production for the market. Some market growers tie paper collars around the celery plants to exclude the light and increase etiolation but this practice is relatively uncommon now because of the high labour requirement.

The trench cultivars were originally developed for autumn and winter use but the season has been considerably extended by the development of the 'self-blanching' types which have gained in popularity. There is an increasing interest in the production of heads or 'sticks' of these latter types as export crops in the USA, especially California, and in some Mediterranean countries, e.g. Israel and Spain. Self-blanching celery has also become an important protected crop for early production in Northern Europe.

In some parts of the Middle East and Asia the plants are grown at a high density and young leaves are cut off and marketed in bunches for flavourings in cooked dishes.

The self-blanching group is not normally earthed-up or blanched by any cultural process except those types that are grown at a relatively high density, and planted rectangularly in beds rather than in rows; these tend to become more etiolated as a result of competition with each other. In addition the European self-blanching cultivars have been developed with relatively little green pigmentation in their petioles. Conversely some of the self-blanching types of American origin have green petioles.

The petioles of self-blanching cultivars of celery are succulent as a result of their wide, fleshy petioles with a high ratio of parenchyma to vascular bundles. It is the strand of collenchyma inside the outer epidermis associated with each vascular bundle which is responsible for the 'stringiness' in celery petioles, the less 'stringy' cultivars having relatively small amounts of collenchyma. The self-blanching types have less frost resistance than the trench types.

Cultivar descriptions of celery

Type for production system:
    Trenching (earthing-up, or collars required for blanching)
    Self-blanching
Petiole characters:
    Colour – White (usually only used where appropriate for trenching cultivars), yellow, green or pink
    Length
    'Stringiness'
    Transverse section of petiole
Resistance to bolting (i.e. flowering in first year)
Resistance to *Septoria apiicola* Speg.

Nutrition and irrigation

Celery requires a soil with a pH of between 6.5 and 7.5 and is not very tolerant of acid conditions. Soils with a high organic matter content are preferable.

Celery has a relatively high moisture requirement and should not be grown in areas of low summer rainfall unless there is an adequate water supply and a satisfactory irrigation system available. Irrigation is especially important in the crop's first year in order to produce heads of a satisfactory size.

The nutrient ratio required by the crop is N : P : K 1 : 2 : 4. Although the market crops (especially summer production of self-blanching types) respond to top dressings of nitrogen these should be applied cautiously for seed production as they will increase susceptibility to frost damage. Some seed producers do, however, apply a top dressing of nitrogen in the spring of the second year.

Flowering

Celery is a biennial, but the market crop is consumed in its first season. All cultivars have a vernalization requirement for flower initiation although a small number of plants with an annual habit occur in some seed stocks. Very little critical work has been done to assess the exact cold requirement for all cultivars, but the self-blanching types tend to bolt readily if young plants are subjected to low temperatures. Hanisova and Krekule (1975) studied methods of flower initiation in cv. Prazsky obrovsky (a Central European cultivar): 20–30 days at 4–6 °C was sufficient to induce bolting and subsequent flowering when plants were lifted in October. A longer period, up to 40 days at the vernalization temperature was required if the plants were lifted in September. They found also that while gibberellic acid did not entirely replace the cold treatment, a 50–100 mg per litre solution of GA_3 on a piece of cotton wool applied to the plant apex between leaf petioles accelerated and increased the uniformity of flowering. This technique of GA_3 application to celery is seen more as a tool for plant breeders to increase the rate of producing successive generations in a breeding programme rather than a technique for commercial seed production.

Bolting can also be induced in the first year if the germinating seed is subjected to temperatures below 10 °C.

After vernalization the plant grows to a height of 1 m in the second year forming a very branched plant with a large number of compound umbels. The individual flowers are relatively small compared with other vegetable seed crops in Umbelliferae.

Pollination

Celery flowers are self-fertile but are largely cross-pollinated by insects. Minimum isolation distances are 500 m. But this isolation distance should be extended between plots or fields for stock seed production, especially between different types.

Agronomy

The plants for seed production are raised and planted out, or are direct sown in the first year for seed production in the second year. If high genetical quality seed is required it is necessary to sow at a similar date to that for the market crop in the area. This allows the seed crop to be grown in the appropriate environment during its early stages.

A later sowing time, after mid-summer (similar to the 'seed to seed' technique for the umbelliferous root crops) is used for the final multiplication stage of commercial seed stock. This still provides the possibility of roguing plants for some vegetative characters such as petiole colour, leaf shape and relative plant height, but it does not allow roguing of plants which would bolt early in the first year of a normal two year crop.

Another factor which must be taken into consideration when deciding on sowing and planting out times is the severity of winter in the proposed seed production area. There is an increasing trend in Europe to sow and plant out at times similar to the commercial market crop and to plant up selected plants in polythene tunnels before winter sets in. It must be emphasized, however, that for stock seed production the crop must be sown in accordance with the requirements for the market crop.

For transplanted crops, the seed is usually sown under protection. There is an increased tendency to sow direct into blocks. The young plants are then hardened-off and planted out when they have developed about five or six leaves.

Plants are generally spaced in the field in rows 60 cm apart, with 30 cm between plants in the rows; these distances can be modified according to the irrigation system used and whether or not the plants will be earthed-up.

In areas with winters of continuous temperatures at or below freezing the plants must be protected. This is usually done by lifting plants in the late autumn and storing them at temperatures just above freezing until re-planting in the spring. Throughout the storage time the roots are kept moist and the foliage relatively dry. If stored in buildings or cellars the relative humidity should be maintained at about 75 per cent. Traditionally, cellar or pit storage has been used and the plants transferred to the field in the spring. This operation is relatively labour intensive and there is an increasing use of polythene structures for selected plants until the seed is harvested.

Harvesting

The seed crop is ready for harvesting when the plants show signs of senility and the majority of seeds on the major inflorescenses have become a grey-brown colour.

Celery seed is notorious for shattering prior to harvest even in the best of conditions and a lot of valuable seed can be lost if bad weather occurs or the cutting time is misjudged.

When produced in tunnels or relatively small areas outside, the plants can be pulled or carefully cut and placed on tarpaulins.

In large-scale production the crop is carefully cut, dried in windrows and combined before excessive shattering has occurred.

If a combine is not used the material can be passed through a stationary thresher, but if dried on tarpaulins, especially under cover, the seed separates from the straw without machine threshing.

Seed is further cleaned by an aspirated screen separator, but because the seed is very small, care must be taken in selection of the bottom screen.

Seed yield and 1000 grain weight

The average seed yield is about 500 kg per hectare. Seed weight: the 1000 grain weight of celery seed is approximately 0.5 g.

Celery seed dormancy

Celery seed normally has a relatively long germination period of up to about three weeks. A further factor which can cause erratic or slow germination is the seed dormancy mechanism which can be due partly to the presence of germination inhibitors in the seed. Thomas *et al.* (1978) found that celery seeds harvested from four umbel positions showed differences in weight and germination characteristics. The primary umbels produced heavier seeds than subsequent umbels but the seeds from primary umbels were less viable when evaluated for germination at 18 °C in the light. They attributed this to the fact that the later the celery seeds are produced on the mother plant, the lower will be the level of endogenous hormones.

Further work by Thomas and O'Toole (1981) concluded that under commercial conditions germination could be improved by soaking the seed in a mixture of gibberellins A_4 and A_7 with ethephon. Their work has important implications in the production of celery seedlings for transplanting, especially where blocking methods of plant propagation are used and in the improvement of germination of pelleted celery seed.

Celery roguing, stages and characters to check

1. *Planting out*   Leaf and petiole characters, plant vigour.

2. *Vegetative stage in field*
(a) Early bolters should be rogued out.
(b) Leaf and petiole characters.
(c) Leaf colour, size, length, colour of leaf tip.
(d) Petioles.
(e) Length.
(f) Colour – light to dark green, yellow, white or pink.

3. *Lifted plants*
(a) Absence of basal shoots (Fig. 13.12).
(b) Width of heart (solidity of plant).
(c) Susceptibility to bolting.

Fig. 13.12   Lifted celery plants at selecting stage, the plant on the right-hand side
has basal shoots and should be rejected.

4. *Second year* (*before flowering*)

(a) Susceptibility to *Septoria apiicola* Speg. and other seed-borne pathogens.

(b) General plant vigour.

For basic seed production, special attention should be given to petiole characters, such as width (transverse section), petiole ribbing and general petiole characters including 'pithiness'.

The main seed-borne celery and celeriac pathogens with common names of the diseases they cause

Pathogens	Common names
Alternaria dauci (Kühn) Groves and Skolko	
Alternaria radicina Meier, Drechsl, and Eddy syn. *Stemphylium radicinum* (Meier, Drechsl, and Eddy) Neergaard	Black mould, root rot
Botrytis cinerea Pers. ex Pers.	Grey mould
Cercospora apii Fresen.	Early blight, leaf spot
Gibberella avenacea Cook syn. *Fusarium avenaceum* (Fr.) Sacc.	
Phoma apiicola Kleb.	Celery root rot, black-neck, scab, seedling canker
Septoria apiicola Speg. syn. *S. apii-graveolentis* Dorogin	Late blight, small leaf spot, large leaf spot
Verticillium albo-atrum Reinke and Berth	
Erwinia carotovora (Jones) Bergey *et al.* syn. *Pectobacterium carotovorum* (Jones) Waldee	Soft rot, crater spot
Pseudomonas apii Jagger	Bacterial blight
Virus	Strawberry latent ring-spot virus

CELERIAC: *Apium graveolens* L. var *rapaceum* (Mill.) DC.

This crop is botanically very similar to celery, but the leaves, especially the petioles are must less significant. The root is thick, approximately the size of a turnip and is used in addition to the leaves in culinary preparations. The root is also used as a salad, especially in France and Belgium.

The cultural requirements and other seed production techniques are the same as for celery, except that the plant spacing is less, usually 60 × 30 cm giving a higher plant population per unit area.

The seed-borne pathogens are the same as for celery. Cultivar descriptions, roguing stages and characters to check are given below.

Cultivar description of celeriac

Season

Use   fresh market, winter storage, processing (dehydration or canning)

Foliage   vigour, intensity of green. Extent to which leaves cover or protect root. Petiole, relative length, colour of petiole bases

Root   relative size and height of root top from soil level, pigmentation, white, purple or red

Resistance to frost, early bolting and *Septoria apiicola* Speg.

Roguing stages and characters for celeriac

1. Vegetative stage, in first season
   Foliage vigour and colour of petioles. Resistance to bolting in first year.
2. Root, relative size, colour, shape with freedom from excessive indentations. Absence of basal shoots. Resistance to seed-borne pathogens, especially *Septoria apiicola* Speg.

REFERENCES

Anon. (1983) *Varieties of Maincrop Carrots*. National Institute of Agricultural Botany, Cambridge.

Austin, R. B. and Longden, P. C. (1966) The effects of manurial treatments on the yield and quality of carrot seed, *J. Hort. Sci.* **41**, 361–370.

Banga, O. (1963) *Main Types of the Western Carotene Carrot and Their Origin*. W. E. J. Tjeenk Willink, Zwolle.

Banga, O. (1964) Identification of Western orange carrot varieties (*Daucus carota* L.), *Proc. Int. Seed Test. Ass.* **29**(4), 957–61.

Bohart, G. E. and Nye, W. P. (1960) *Insect Pollinators of Carrots in Utah*. Bulletin 419, Agriculture Experiment Station, Utah State University.

Cooke, G. W. (1975) *Fertilizing for Maximum Yield*. Granada Publishing.

Gray, D. (1979) The germination response to temperature of carrot seeds from different umbels and times of harvest of the seed crop, *Seed Sci. Technol.* **7**, 169–78.

Gray, D. (1981) Are the plant densities currently used for carrot seed production too low? *Acta Horticulturae* **111**, 159–65.

Gray, D. and Steckel, Joyce A. (1980) Studies on the sources of variation in plant weight in *Daucus carota* (carrot) and the implications for seed production techniques. In *Seed Production*, P. D. Hebblethwaite (ed.), pp. 474–84. Butterworths, London.

Hanisova, A. and Krekule, J. (1975) Treatments to shorten the development period of celery. *J. Hort. Sci.* **50**, 97–104.

Hawthorn, L. R. (1951) *Studies on Soil Moisture and Spacing for Seed Crops of Carrots and Onions*. USDA, Circular No. 892.

Hawthorn, L. R. and Pollard, L. H. (1954) *Vegetable and Flower Seed Production.* The Blakiston Company Inc., New York.

Hawthorn, L. R., Bohart, G. E., Toole, E. H., Nye, W. P. and Levin, M. D. (1960) *Carrot Seed Production as Affected by Insect Pollination,* Bulletin 422, Agriculture Experiment Station, Utah State University.

Jacobsohn, R. and Globerson, D. (1980) *Daucus carota* (carrot) seed quality: II The importance of the primary umbel in carrot seed production. In *Seed Production,* P. D. Hebblethwaite (ed.). pp. 637–46. Butterworths, London–Boston.

Sandin, N. H. (1980) Optimum harvest time for *Daucus carota* (carrot) seed crops in Sweden. In *Seed Production,* P. D. Hebblethwaite (ed.)., pp. 553–9. Butterworths, London–Boston.

Thomas, T. H. and O'Toole, Diane F. (1981) Environment and chemical effects on celery (*Apium graveolens* L.) seed production, *Acta Horticulturae* **111**, 131–8.

Thomas, T. H., Gray, D. and Biddington, N. L. (1978) The influence of the position of the seed on the mother plant on seed and seedling emergence, *Acta Horticulturae* **83**, 57–66.

14 ALLIACEAE

The only genus in this family of importance to vegetable producers is *Allium* and although there are a large number of cultivated species in this genus the following are the most important.

Allium cepa L.	Onion
Allium ampeloprasum L. var *porrum*	Leek
(syn. *A. porrum* (L). Gay)	
Allium ascalonicum L.	Shallot
Allium fistulosum L.	Japanese bunching or Welsh onion
Allium sativum L.	Garlic
Allium schoenoprasum L.	Chives
Allium tuberosum Rottl. ex Spreng	Chinese chives

The two most important of the above crops are onion and leek which are normally grown from seed; the remainder are mainly vegetatively propagated although occasionally grown from seed.

ONION: *Allium cepa* L.

Allium cepa L. probably originates in Afghanistan, Iran and Pakistan but is now cultivated in many areas of the world including the tropics and the temperate regions. The storage potential of the mature bulbs has led to onions becoming an important crop. The very young plants, developing and mature bulbs are used in a variety of ways. There are several uses for onions in processing including, pickling, chutneys, sauces and dehydration. The ranges of cultivation areas and uses have led to a large number of cultivars and types.

As onion bulb formation is dependent on daylength, there are specific daylength groups for different latitudes, from types with a sixteen hour daylength requirement adapted to the northern and southern latitudes to twelve hour daylength types suited for the tropics. This photoperiodic effect on bulb formation is discussed by Jones and Mann (1963).

Cultivar description of onion

Method of seed production	Open pollinated or hybrid
Specific uses	salad (bunching), bulbing

Season early, maincrop, suitability for storage.
Suitability for sowing at specific times of
year, e.g. autumn (plants capable of overwintering) or spring sowing

Cultivar suitable for specific market outlet or processing (e.g. scale colour
and dry matter content is of interest to dehydrators who prefer white internal bulbs with a relatively high dry matter content)

Photoperiod for bulbing

Bulb characters

Shape: flat, globe or cylindrical. (There are also flat top and high top globe
types.)
Outer skin colour: white, yellow, brown, pink, red or green.
Skin quality when ripe in relation to thickness and numbers of ripe dry (i.e.
ripe) protective layers of skin.
Flesh colour.
Pungency, relative pungency.

Leaf characters, colour and pose (Fig. 14.1).

Resistance to specific pathogens, e.g. *Pyrenochaeta terrestris* (Hans.) Gorenz,
J. C. Walker and Larson (pink root), and *Perenospora destructor* (Berk.) Casp.
(downy mildew).

Fig. 14.1 Onion cultivars showing different foliage characters.

Agronomy

pH and nutrition

Onions will grow satisfactorily on soils with a pH of 6.0–6.8.

The ratios of N : P : K applied during seed bed production are 1 : 2 : 2 although some bulb producers increase the nitrogen ratio according to the soil status. Very little work has been reported on the effects of nutrition in the first year on seed produced in the second year. Work by Ahmed (1982) has shown that the nutrition of the bulbs in their first year can have an important effect on yield and quality of seed in the second year. He found that nitrogen applications equivalent to 150 kg/ha combined with similar levels of P and K produced plants which yield the largest bulbs and highest total bulb yield at the end of the first year. Ahmed continued his mineral nutrition experiments with bulbs produced under known nutrient regimes. After storage and re-planting he found that seed yields were increased by further applications of nitrogen up to levels equivalent to 150 kg/ha whereas further P and K had no effect unless nitrogen was applied. However, an extremely important finding from Ahmed's work was that whereas the supplementary nitrogen in the second year increased seed yield, the seed quality (as measured by germination tests and an accelerated ageing test) was reduced. Ahmed reported that top dressings of N during flowering at rates not exceeding 100 kg/ha enhanced the production of good quality seed without severely reducing seed yield. This is important as onion seed is both a relatively valuable commodity and the seeds have a relatively short shelf-life when compared with many other vegetable seeds.

Mother plants which had received relatively high potassium levels in their first year had proportionately higher potassium levels in the mature bulb and produced seed of high quality in their second year. This demonstrated the carry-over effect of nutrient regimes from the year of bulb production to the second year of seed production.

Irrigation

Hawthorn (1951) found that high soil moisture in the seeding year favoured high seed yield. This was especially so at high plant densities with rows less than 30 in apart (approximately 76 cm). Work by Millar *et al.* (1971) demonstrated that there is a large drop in water potential between the flowers and upper part of the seed stalk. They also reported lesser, but still important, differences between water potential of flowers and other parts of the plant. This work clearly demonstrated the moisture-sensitive stage of onions throughout anthesis.

In practice, the soil surface should not be continuously wet because it will predispose the crop to infection by *Botrytis allii* Munn (neck rot or damping off). Bulbs generally take longer to ripen in wet seasons or if unnecessary irrigation water has been applied towards the end of the season.

Crop husbandry

The 'seed to seed' and 'bulb to seed' methods are both used in onion seed production. The former does not allow for bulb inspection as relatively small plants overwinter from a summer sowing and the method is not suitable for basic seed production.

The bulb to seed method is a longer process, with mature bulbs produced by the end of the first season (as in the commercial production of bulbs for market). This allows roguing and selection to include mature bulb morphology as this is especially important for basic seed production. The selected bulbs flower and seed in their second year.

Seed to seed

The seeds are drilled in mid-summer to early autumn, depending on the cultivar and local climate. Generally the sowing date is towards the end of this period in warmer areas. The plants have to reach sufficient size for vernalization. Seed is sown at the rate of 4–5 kg/ha in rows 70–100 cm apart. Although the 'seed to seed' crop cannot be examined for mature bulb characters the fields are rogued at the end of the summer to remove obvious off-types, e.g. incorrectly coloured bulbs.

Bulb to seed

The seeds for this method are sown either in single rows on the flat 40–55 cm apart or in beds with a distance of 90–100 cm between bed centres and rows within the beds 30–40 cm apart. A sowing rate of 3–6 kg/ha is higher than that adopted for the production of a commercial bulb crop for market because a smaller bulb of approximately 5–8 cm diameter is preferred. One hectare of bulbs from the first year will plant up 5 ha for the seed production.

When the tops die down at the end of the first growing season the bulbs are lifted and dried. In some areas the tops are trimmed off with a knife but Bleasdale and Thompson (1965) have shown that it is better for the tops to dry off naturally and to keep the bulbs aerated during this 'curing' period than to accelerate the process by trimming. The bulbs are rogued after curing (Figs. 14.2 and 14.3), stored (see below) and examined again before re-planting the following season. In some areas the bulbs are re-planted immediately after curing and sorting but this can be done only when winter conditions are suitable (either relatively mild or a covering of snow protects the over-wintering bulbs). Mother bulbs are re-planted in furrows 70–100 cm apart (Fig. 14.4).

Flowering

Flower initiation

The vernalization requirement of onion plants for satisfactory flower initiation varies according to the cultivar. Those used in temperate regions generally have a relatively long vernalization requirement, whereas those developed in the tropics have a very short vernalization period. Sinnadurai (1970) reported a Nigerian cultivar which would initiate flowers with no low temperature treatment. The author has also found local cultivars in the Sudan

Fig. 14.2   Sorting and roguing cured onion bulbs in the Sudan prior to storage.

Fig. 14.3   Range of bulb qualities found when sorting cured bulbs, bottom two
rows satisfactory, centre mechanically damaged, top left group 'bull-
necks' and top right 'splits'.

Fig. 14.4   Mother onion bulbs re-planted after curing and sorting, Golan Heights, Israel.

which have no low temperature requirement for flower initiation (George 1980).

Bulbs produced for the bulb to seed method either receive their cold stimulus in store or in the soil if re-planted immediately after selection. It is especially important to ensure that stored bulbs receive sufficient cold stimulus for flower initiation. Brewster (1977b) reviewed the available evidence and concluded that the most favourable storage temperature for flower initiation was 9–13 °C. The storage of mother bulbs in huts or other structures in the tropics before re-planting is more closely related to avoiding bulb loss from extremely high temperatures than the provision of a lower temperature for vernalization (Fig. 14.5).

Fig. 14.5 Traditional straw hut or 'cottage' used for onion bulb storage in the Sudan.

Plants grown by the 'seed to seed' method overwinter in the field and therefore receive their cold stimulus *in situ*. Cultivars which overwinter from an autumn sowing for normal market production the following year and the newer hybrids developed in Japan which are suitable for autumn sowing do not normally bolt because the plants are not large enough to receive sufficient cold stimulus.

Flower development and pollination

The onion flowers are borne on umbels. The number of umbels per plant depends on several factors including storage environment of the bulb, plant density and cultivar. Ahmed (1982) found that nutrient regime in the first year did not affect numbers of seed stalks (inflorescences) per plant although he reported that a high nitrogen regime in the first year of bulb production significantly increased seed stalk height in the second year, which could lead to increased lodging.

Pollination

The duration of anthesis is approximately four weeks on individual umbels and there is a complicated succession of flowers opening on each. The sequence of flower opening and incidence of protandry in onion flowers has been studied by Currah and Ockendon (1978). The effects of environmental factors on onion pollen and ovule development have been extensively reviewed by Currah (1981).

The flowers are pollinated by bees, flies and other insects. The production of F_1 hybrid onion seed and problems associated with pollination and insect behaviour, including the insufficient number of insect visits between male fertile and female plants, has stimulated several investigations, also reviewed by Currah (1981).

Isolation

The minimum recommended isolation distance between different cultivars is 1000 m. Some authorities stipulate shorter distances than this for cultivars with the same bulb colour. In some countries there are declared zones in which only cultivars of a specific bulb colour can be grown for seed.

Roguing stages

Seed to seed

1. *During autumn of first season*   Remove plants with off-type foliage, off-type bulb or stem colour and plants bolting in first year.

2. *Early flowering in second year*   Remove plants with off-type foliage, off-type bulb or stem colour and check inflorescence characters where appropriate.

Bulb to seed

1. *Before bulb maturity*   Remove plants with off-type foliage, off-type bulb or stem colour, plants bolting in first year and late maturing plants.

2. *When sorting lifted bulbs*   Check that bulb shape, colour and relative size are true to type. Discard early bolters, bull necks, bottle-shaped bulbs, split bulbs, doubles, damaged and diseased bulbs (Figs 14.3 and 14.6).

3. *At re-planting*   Check characters as described above when sorting and also discard early sprouting bulbs.

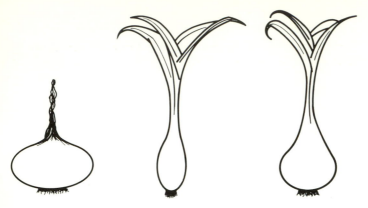

Fig. 14.6   Onion bulbs at end of first season: left satisfactory bulb with thin neck, middle 'bull-neck' and right bottle-shaped bulb.

4. *At start of flowering*   When appropriate check inflorescence and flower characters. Flowering heads of plants infected with *Ditylenchus dipsaci* (Kühn) Filipjev tend to be bent over; infected plants should be removed and burnt.

Production of F₁ hybrid onion seed

The techniques used in the maintenance of parental lines have been described by Watts (1980). The usual ratio of male to female rows is 1 : 4 or 1 : 8. The pattern and ratio depend upon mechanization and also upon the amount of pollen produced by the male parent. In some cases where there is difficulty in synchronized flowering of the two parents, differential storage treatments are used for the bulb-lots of the two parents. Either parent may require roguing according to plant breeders' instructions to remove off-types (e.g. plants with male fertile flowers in a male sterile female parent).

Seeds produced on the male parents are either harvested or the rows are rotavated according to the value of seed before harvesting from the female parent commences. Care must also be taken to ensure that one or other parent is clearly marked at the ends of the rows as morphological differences are sometimes not obvious. Some seed producers paint the end few plants on each female row.

Basic seed production

Basic onion seed is produced only by the 'bulb to seed' system. Mother bulbs are usually subjected to more critical criteria than for commercial seed. Bulb hardness is sometimes an additional criterion, although this is more appropriate in selecting material in breeding programmes (Fennell 1978).

Seed-borne pathogens

The main seed-borne pathogens of onions with the common names of the diseases they cause are listed below under *Allium* spp. at the end of the section on leeks.

The use of growth-regulating substances in onion seed production

Investigations by Thomas (1969) showed their importance in extending bulb dormancy, and the results of Van Kampen and Wiebosch (1970) demonstrated that seed yield can be increased by soaking the bulbs in solutions of growth regulators.

Levy *et al.* (1972) investigated the effects of ethephon on the growth of seed stalks (scapes) and seed yield. They reported that two field applications of ethephon at 480 ppm starting when 75 per cent of the plants had visible seed stalks reduced seed stalk height without affecting seed yield or germination. This effect has the advantage of reducing lodging and facilitates mechanical harvesting. Later work by Naamni *et al.* (1980) investigated the treatment of mother bulbs with GA_3 when the first flower stems were emerging: a single application of GA_3 at 50 ppm reduced the time to 80 per cent floral stem emergence by a half and improved the uniformity of seed stalk height (Fig. 14.7).

Plants treated with GA_3 had larger umbels with a 30 per cent increase in seed yield but with no reduction in seed germination (Table 14.1).

Harvesting

Traditionally onion seed heads have been harvested by hand when approximately 5 per cent of the capsules on individual heads are shedding ripe seeds. The seeds are black when ripe and can be seen against the silvery coloured capsules. The seed heads shatter readily and the exact timing of this operation is based on experience and local weather.

The seed heads with approximately 10–20 cm of scape (seed stalk) attached are removed by cutting with a sharp knife or secateurs. When cutting, the umbel is supported in the palm of the hand and held between the fingers to avoid loss of seed. In some onion seed production areas the crop is cut once over, but in others several successive hand harvests are made.

The seed heads are further dried on tarpaulins or sheets either in the open or in suitable structures. Some producers have a tiered box or crate system to allow air to circulate freely.

There is a great deal of interest in the development of mechanical harvesters for this crop. In North America successive specialist onion seed producers have been developing their production methods for nearly a hundred years. During the last two decades attempts have been made to mechanize the cutting operation. One system is to cut the stems at approximately 15 cm above

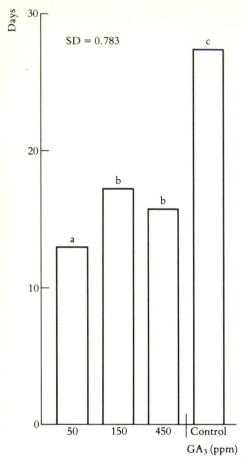

Fig. 14.7   Effects of GA$_3$ concentration on number of days to 80 per cent onion
seedstalk emergence. Histograms not headed by the same letter are
significantly different at the 5 per cent level (from Naamni *et al.* 1980).

ground level with mowers and pick up the cut material with an elevator
which places it in mobile containers. The material is then further dried on
sheets. The presence of the long stems in the heaps increases the incidence
of the material heating up especially if the heaps are up to 200 cm deep, in
which case they must be turned every five or six days.

Another move towards mechanization is the system of mechanically cutting two rows at a time and drying in windrows which are placed on
machine–laid paper in the former irrigation furrow between two rows. When
the material is sufficiently dry both it and the paper are picked up by a combine, but despite picking up the paper containing loose seeds the method is
wasteful and unlikely to be developed further. Other developments include
modified combines with the threshing cylinder removed.

Work in Israel has shown that direct harvesting and threshing of the onion
seed crop lead to relatively high losses. As a result of this Globerson *et al.*
(1981) examined the nature of flowering and seed maturation in relation to

Table 14.1 The effects of GA₃ application on onion seed yield, 1000 grain weight and partial germination (from Naamni et al., 1980)

GA₃ treatment (ppm)	Seed yield (g/10 m²)	Seed distribution			Thousand weight of seeds		Seed germination (%)	
		From large umbels (%)	Wt per large umbel (g)	Wt per small umbel (g)	Large umbels (g)	Small umbels (g)	Large umbels	Small umbels
0	911.8[b†]	81.6[b]	3.28	1.32[b]	3.49[b]	3.24	92.80	88.40
50	1322.2[a]	91.4[a]	3.43	1.53[b]	3.58[ab]	3.13	92.20	92.50
150	1044.9[b]	88.6[a]	3.04	1.70[a]	3.69[ab]	3.28	91.10	90.60
450	1093.8[b]	89.9[a]	3.06	1.32[b]	3.91[a]	3.29	92.40	85.10
SE	62.2	0.9	0.12NS[‡]	0.08	0.10	0.12NS	2.21NS	2.35NS

* Mean values of treatments for both dates of GA₃ application.
† Values within a column not followed by the same letter are significantly different at the 5% level.
‡ Not significant

mechanical harvesting: the best time for mechanical harvesting was when the seeds had a dry matter content of 60–70 per cent; seeds which were dried while still in capsules attached to the stalks germinated better than those dried in their capsules after separation from the umbels. A larger quantity of properly mature seed can be obtained if shattering is prevented, and further work is in progress at several centres in the world to develop the use of anti-shattering materials as discussed in Chapter 4.

Threshing and cleaning

The material is ready for threshing as soon as it is dry and the seeds can be separated from their capsules by rubbing in the hand. In order to avoid damaging the brittle seeds threshing should not be delayed beyond this stage. Several threshing methods have been adopted for onions depending on the scale of operation and include flailing, rolling, threshing machines and combines. Onion seeds are very easily damaged during processing and frequent checks should be made to ensure that the seed coats are not accidently cracked during any operation. Examination of samples with a hand lens will confirm if the processing is satisfactory. Concave settings should be adjusted to avoid injury to the seed. Another important point is to ensure that the processing does not break too many of the former flower pedicels from their stalks as these are difficult to separate from the seed-lot in subsequent cleaning processes. The light debris ('tailings') must also be examined occasionally to ensure that it does not contain an undue proportion of good seeds.

After threshing, the initial cleaning is usually achieved with an air-screen machine. Further upgrading according to the state of the seed-lot can be done by passing the seeds over gravity tables. Seed-lots with unacceptable levels of flower pedicels remaining can be further upgraded either by using a magnetic separator (the iron powder adheres to the pedicels and not the seeds) or by flotation. But in the latter process it is usual practice to put the light fraction only into water. During this process, which should not exceed three minutes per seed-lot, the good seeds sink while the poor quality light seeds and pedicels float off, the heavier debris absorbs water and swells. After spin drying and further drying in drying racks the large debris is removed by an air-screen cleaner. The final onion seed-lot must then be dried down to a moisture content not exceeding 12 per cent or lower depending on method of storage and packaging (see Chapter 5).

Seed yield

The best yield from open-pollinated crops produced under ideal conditions is *c.* 2000 kg/ha but *c.* 1000 kg/ha is more common and in some areas yields of *c.* 500 kg/ha are frequently accepted as satisfactory.

The yield from F_1 hybrids is normally lower than from open-pollinated crops and is often as low as *c.* 50–100 kg/ha.

Seed weight

The 1000 grain weight for onions is *c.* 3.6 g.

LEEK: *Allium ampeloprasum* L. var. *porrum*

Leeks are believed to originate from the eastern Mediterranean and are now widely cultivated in temperate regions. The mature plants are especially important as a winter vegetable in Northern Europe and in the Middle East and Asia the young plants are cut and used in salads or stews.

There is relatively little diversity of leek types compared with most other vegetable crops.

Cultivar description of leek

Maturity period, resistance to frost, resistance to early bolting

Height of shaft or column (i.e. the blanched petioles), relative thickness of shaft

Degree of bulbing at the shaft base

Leaf characters   (The foliage above soil level is sometimes referred to as the 'flag'.)

Colour: green, grey-green, tendency to become blue-green during periods of low temperature
Shape, including relative length and width: the leaves of some cultivars have a characteristic 'keel' to the leaf

Agronomy

Soil pH and nutrition

No experimental work relating to soil and nutrient requirements of leeks for seed production has been reported in the literature. However, the market crop is known to respond to soils with a high organic matter content and dressings of up to 60 tonnes/ha of bulky organic manures. These are applied as farmyard or other suitable material during winter preparations.

Leeks tolerate a slightly acid soil and a pH between 6.0 and 6.8 is satisfactory. Calcareous materials to increase the soil pH should not be applied at the same time as bulky organic manures. The ratio of N : P : K applied during the final stages of soil preparations is 2 : 2 : 1, but the nutrient values of any bulky organic manures which have been applied must also be taken into account when calculating the amounts of fertilizers to apply.

Plant production

Leeks are biennials and normal flowering occurs in the second year after vernalization. The plants are not stored during the winter but remain in the ground. Both 'seed to seed' and 'root to seed' systems are used for leek seed production but only 'root to seed' is used for the production of basic seed.

Seed to seed method
The seeds are drilled in rows 30 cm apart at a rate of 2–3 kg/ha. At this sowing rate there is no need for subsequent thinning if the seeds are sown evenly. If higher rates are used then it is necessary to thin the young plants, as soon as they are large enough to handle, to approximately 10 cm apart in the rows. The 'seed to seed' system can present weed control problems and for this reason alone it is unpopular with some leek seed producers.

Root to seed method
This is the only method used for basic seed production and is often preferred for the production of commercial seed stocks although it involves extra labour and costs for the transplanting operation.

Seeds are sown in early spring in seed beds at the rate of 4–5 kg/ha. After eight to twelve weeks the seedlings are lifted and graded. Some producers trim the plants during grading. Plants which are approximately pencil thickness are re-planted in 10 cm deep prepared drills 8–10 cm apart with 70 cm between the rows. The transplants are either dibbled in or planted by machine.

Adamson (1952), working in Southern Vancouver Island, Canada, investigated time of sowing and transplanting: when field sowings were made by 15 May there was no significant reduction in seed yield although there was a tendency for the earliest sowings to produce more seeds. He found that a close spacing of 2 in (*c.* 5 cm) within the rows for direct drilled crops gave a greater total seed yield per unit area than a 6 in (*c.* 15 cm) spacing although the higher plant density gave a lower seed yield per plant. When Adamson compared the seed yield from a July transplanting at 6 in (*c.* 15 cm) within the rows with a spring field sowing which was thinned to the same distance within the rows, he found that the seed yields were similar and that there was no marked differences in seed germination although he did not report on other aspects of seed quality; relatively late transplantings (i.e. in the autumn) from seed beds sown in early June resulted in low seed yields. It is generally recommended that a crop for seed should be sown at approximately the same time as a market crop would be and that late transplanting is avoided.

Leek roguing stages and main characters to be observed

'Seed to seed' method

1. *In summer of the first season*   When plants are well developed, check leaf colour, relative leaf size, width and keel characters; presence or absence of pigmentation according to the type.

2. *At start of flowering*   As above and also check colour of inflorescences.

'Root to seed' method

1. *When transplanting from seed beds*   Check leaf colour, relative size, width and keel characters; also that presence or absence of stem pigmentation is according to the type.

2. *In late summer*   After transplants are established and actively growing check as described above for stage 1. Also remove any plants which are bolting in their first year and check that vigour of individual plants is according to the type.

3. *When the crop starts to flower in the second year*   As described above for stage 1. Also check inflorescence colour.

Flowering, pollination and isolation

These, including isolation distances, are similar to onion although there are normally no special zoning arrangements.

Basic seed production

The 'root to seed' method described above is used but the young plants are 'dibbled' individually at transplanting time. During the late autumn or early winter the mature plants are lifted and selected for trueness to type. Particular attention is given to stem quality and freedom from 'bulbing' at the stem base, according to the cultivar's characteristics. The selected plants are immediately re-planted, preferably into polythene tunnels (Fig. 14.8).

Fig. 14.8   Basic seed production of leeks; planting selected material for seeding in a polythene tunnel.

Harvesting, drying, seed extraction and processing

Leek seeds are harvested by hand as described for onions. Leeks tend to flower later than onions and as their seed heads are also later ripening care must be taken not to lose seed as a result of inclement autumn weather. Seed heads may require artificial drying after harvesting. Adamson (1960) investigated the effects on germination of drying seed heads at different temperatures. He reported that material which had been dried at a continuous temperature of 80 °F (26.7 °C) resulted in 65 per cent or higher germination. He also found that seeds produced in a relatively cool summer (as determined by heat unit data) and dried at relatively high temperatures were more readily adversely affected by continuous drying treatments of 95 °F (35.0 °C) and combinations of 80 °F (26.7 °C) and 90 °F (32.2 °C).

The extraction and processing of leek seeds is the same as described for onion seeds.

Leek seed yield and 1000 grain weight

The yield is *c.* 500 kg/ha although some producers achieve yields of up to 600 kg/ha.

The 1000 grain weight is *c.* 3.75 g.

The main seed-borne pathogens of *Allium spp.* with common names of the diseases they cause

Pathogens	Common names
Alternaria porri (Ell.) Ciferri	Purple blotch
Botrytis allii Munn	Damping-off, grey mould, neck rot
Botrytis byssoidea Walker	Seedling damping-off, neck rot
Cladosporium allii-cepae (Ranojević) M. B. Ellis syn. *Heterosporium allii-cepae* Ranojević	
Colletotrichum circinans (Berk.) Vogl., syns. *C. dematium* (Pers. ex Fr.) Grove *F. circinans* (Berk.) Arx	Smudge, damping-off
Fusarium spp.	
Perenospora destructor (Berk.) Casp.	Downy mildew
Pleospora herbarum (Pers. ex Fr.) Rabenh., syn. *Stemphylium botryosum* Wallr.	Black stalk rot, Leaf mould
Puccinia allii Rud., syn. *Puccinia porri* Wint.	Rust
Sclerotium cepivorum Berk.	White rot
Urocystis cepulae Frost	Smut
Virus	Onion yellow dwarf virus
Ditylenchus dipsaci (Kühn) Filipjev	Bloat, eelworm rot

REFERENCES

Adamson, R. M. (1952) Effects of various growing methods in leek seed production, *Scientific Agriculture* **32**, 634–7.

Adamson, R. M. (1960) The effect of germination of drying leek seed heads at different temperatures, *Canad. J. Plant Sci.* **40** 666–71.

Ahmed, A. A. (1982) The influence of mineral nutrition on seed yield and quality of onion (*Allium cepa* L.). Thesis submitted for the degree of Doctor of Philosophy, University of Bath.

Bleasdale, J. K. A. and Thompson, R. (1965) Onion skin colour and keeping quality. *Rep. Nat. Veg. Res. Sta* **16**, 47–9.

Brewster (1977a) *see* 'Further Reading', *below*

Currah, L. (1981) Onion flowering and seed production, *Sci. Hort.* **32**, 26–46.

Currah, L. and Ockendon, D. J. (1978) Protandry and the sequence of flower opening in the onion (*Allium cepa* L.), *New Phytologist* **81**, 419–28.

Fennell, J. F. M. (1978) Use of a durometer to assess onion bulb hardness, *Exp. Agric.*, **14**, 169–72.

George, R. A. T. (1980) Onion germplasm in the Sudan, *Plant Genetic Resources*, **42**, 18–20. FAO/International Board for Plant Genetic Resources.

Globerson, D., Sharir, A. and Eliasi, R. (1981) The nature of flowering and seed maturation of onions as a basis for mechanical harvesting of the seeds, *Acta Horticulturae* **111**, 99–114.

Hawthorn, L. R. (1951) *Studies of Soil Moisture and Spacing for Seed Crops of Carrots and Onions.* Circular 892, USDA, Washington DC.

Jones, H. A. and Mann, L. K. (1963) *Onions and Their Allies.* Leonard Hill, London. Interscience Publishers, New York.

Levy et al. (1972) *see* 'Further Reading' *below*.

Millar, A. A., Gardner, W. R. and Goltz, S. M. (1971) Internal water status and water transport in seed onion plants, *Agron. J.* **63**, 779–84.

Naamni, F., Rabinowitch, H. D. and Kedar N. (1980) The effect of GA$_3$ application on flowering and seed production in onion, *J. Amer. Soc. Hort. Sci.* **105** (2), 164–7.

Sinnadurai, S. (1970) The effect of light and temperature on onions, *Ghana J. Agric. Sci.* **3**, 13–15.

Thomas, T. H. (1969) The role of growth substances in the regulation of onion bulb dormancy, *J. Exp. Bot.*, **20**, 124–37.

Van Kampen, J. and Wiebosch, W. A. (1970) Experiments with some growth regulators for seed production in onions, *Medelingen Proefstation voor de Groenteteelf in de Vollegrond*, No. 47.

Watts, L. E. (1980) *Flower and Vegetable Plant Breeding.* Grower Books, London.

FURTHER READING

Astley, D., Innes, N. L. and van der Meer, Q. P. (1982) *Genetic Resources of Allium Species.* International Board for Plant Genetic Resources, FAO, Rome.

Brewster, J. L. (1977a) The physiology of the onion (Part 1), *Hort. Abstr.* **47**(1), 17–23.

Brewster, J. L. (1977b) The physiology of the onion (Part 2), *Hort. Abstr.* **47**(2), 103–12.

Levy, D., Ventura, J. and Kedar, N. (1972) The effect of ethephon on seed stalk growth and seed yield of onion, *Hortscience* **7**(5), 470–1.

Woike, H. (1981) Some aspects of the role of the honey-bee in onion seed production in Poland, *Acta Horticulturae*, **111**, 91–8.

15 GRAMINEAE

This is considered to be the most important flowering plant family, containing many genera including cereals, fodder grasses, sugar cane, sorghums, millets and bamboos. Most cultivated crops in this family are of agricultural importance, but the only ones considered as vegetables are some vegetatively propagated bamboos in Asia and the sweet corn group within the wide range of maize types (*Zea mays* L.).

ZEA MAYS L.: Sweet Corn, Corn on the Cob, Vegetable Corn

According to Purseglove (1972) the cultivars of *Zea mays* can be classified into seven groups.

1. Pod corn – *Zea mays tunica* Sturt.
2. Popcorn – *Zea mays everata* Sturt. (syn. praecox),
3. Flint maize – *Zea mays indurata* Sturt.
4. Dent maize – *Zea mays indenata* Sturt.
5. Soft or flour maize – *Zea mays amylacea* Sturt. (syn. *erythrolepis*)
6. Sweet corn – *Zea mays saccharata* Sturt. (syn. *sugosa* Bonof.)
7. Waxy maize – *Zea mays ceritina* Kulesh.

There are also ornamental forms, e.g. var. *japonica* Koern. which has striped leaves and var. *gracillima* Koern. which is a dwarf form. The different cultivars of all these groups cross-pollinate and it is sometimes difficult to classify them exactly. The only group considered important as a vegetable is *Zea mays* L. *saccharata* Sturt. – the sweet corn cultivars which have a higher proportion of sugars to starches in the seeds' endosperm than the dent maize which is the main type grown for fodder and dry grain. Within the sweet corn group there are supersweet cultivars with a genetically controlled factor for extra sugar in the seeds which takes longer to change to starch after harvesting; they are therefore sweeter at harvest and remain sweeter post-harvest. Hybrid cultivars of sweet corn are widely adopted in Europe and North America but open-pollinated cultivars are still used in Asia.

Cultivar description of sweet corn

Type of cultivar 'normal' or 'supersweet'

Season of use and suitability for specific outlets (e.g. fresh market, canning or freezing)

Plant height	relative height of plant when pollination occurs
Inflorescences	general colour of male inflorescences ('tassel'), general colour of female inflorescence ('silk')
Characters of cob (sometimes referred to as 'ears')	length and shape, approximate number of rows of kernels (i.e. immature seeds) colour of kernels at market maturity
Method of cultivar maintenance	e.g. open-pollinated, F_1 hybrid, synthetic hybrid

Agronomy

Soil, pH and crop nutrition

Sweet corn tolerates a range of soil types, but it is important that the soil should have good water retention characters. The crop tolerates a pH between 5.5 and 6.8 and appropriate quantities of lime should be applied if the pH is below this. N : P : K fertilizer application during final stages of ground preparation should be in the ratio of 1 : 2 : 2. Some growers apply a higher proportion of nitrogen in the base dressing while others give a nitrogenous top dressing during the young plant stage.

Sowing and spacing

The sowing rate is regulated according to the relative seed size of the cultivar and seed-lot. The larger seeded cultivars are generally sown at a rate of up to 30 kg/ha and the smaller seeded types at 15 kg/ha in rows 70–100 cm apart depending on machinery used for cultivations and other operations.

Irrigation

As sweet corn seed is produced in areas with relatively low summer rainfall it is important to ensure that sufficient irrigation is available. MacKay and Eaves (1962) found that sweet corn was very responsive to supplementary irrigation especially from the pollen-shedding stage to cob maturity; also that removing the moisture stress by increasing the irrigation increased the crop's response to nitrogen, phosphorus and potassium; they attributed this to a response of larger plants resulting from the irrigation. Although there has been relatively little quantitative work done on sweet corn seed production

compared with that on maize, the evidence for sweet corn response to irrigation is generally in accord with the findings of workers with maize in that
there is a critical period during and immediately post-anthesis when soil
moisture should be maintained.

Flowering, pollination and isolation

The male flowers ('tassels') are in a terminal infloresence and the female
flowers ('cobs' or 'ears') are borne on lateral branches (Fig. 15.1). There is a
tendency for dehiscence of pollen up to two or three days in advance of
the stigmas becoming receptive, which increases the incidence of cross-
pollination.

The pollen is wind-borne and isolation distances of up to 1 km are necess-

Male
flowers

Female
flowers

Fig. 15.1   Plants of sweet corn showing positions of male flowers ('tassels') and
female flowers ('cobs' or 'ears').

ary. It is extremely important to ensure maximum isolation between a sweet corn seed crop and other crops of *Zea mays* grown for any purpose, including fodder. Attention should be given to checking that volunteer *Zea mays* plants are removed from the isolation area. Some seed producers increase the number of perimeter rows of male parents to reduce the proportion of 'foreign' wind-borne pollen entering the crop.

Roguing

Both parents must be inspected if hybrid seed is being produced and roguing criteria applied according to the description of each parent line. Roguing is not normally done in stages but attention must also be given to detailed descriptions provided by plant breeders of likely off-types. The roguing operations must be completed before male or female inflorescences are fully developed.

An additional check for trueness to type (especially characters such as seed colour) can be made after harvesting when the husks are removed from the cobs.

Production of hybrid seed

Parent inbred lines for hybrid seed production are normally maintained by, or under the supervision of, plant breeders. The ratio of female to male parent rows is based on the quantity of pollen produced by the male parent, but is generally in the proportion of either 2 : 1 or 4 : 2 female to male respectively. Unless there is satisfactory genetical or chemical control of pollen production the male inflorescences ('tassels') are either removed by hand or mechanically. This operation, known as 'de-tasselling', is done before the anthers dehisce; male inflorescences must also be removed from tillers. It may be necessary to de-tassel the crop up to seven times, at two day intervals, to ensure that it is complete. Specialist hybrid sweet corn seed producers in the USA have developed propelled platforms upon which the workers (known as 'de-tassellers') stand during this operation.

Harvesting and processing

Clear instructions must be given to harvesting gangs regarding the disposal of seeds from male lines before the main harvest commences. The cobs of the main seed crop are harvested when the seeds' moisture content is down to 45 per cent or slightly less. Visual signs of this stage are the glazed appearance of seeds as they harden. Small areas are harvested by hand but specialists in the USA either use mechanical pickers or direct combining. Further post-harvest drying of cobs is done in bird- and rodent-proof cages known as 'cribs', or the cobs are dried artificially.

Seeds are not removed from cobs (usually called 'shelling') until the seeds' moisture content has reduced to 12 per cent. The husks have to be removed from the cobs before the shelling operation which also provides a post-harvest check of genetical quality and an opportunity to discard cobs displaying diseased seeds.

The shelling is done by a maize or corn sheller, a machine which pushes the cobs on to a cylinder with peg-like projections, separating the seeds from the cob's fibrous axis.

After shelling the seeds are passed through an air-screen cleaner and finally graded.

Seed yields and 1000 grain weight

The seed yield depends on the cultivar and, if hybrid seed is produced, the proportion of plants cultivated as the female parent. Yields of 1500 kg/ha are generally obtained although specialist producers in the USA obtain up to 2500 kg/ha.

The 1000 grain weight of the super sweet cultivars is *c.* 100 g and is *c.* 210 g for 'normal' cultivars.

The main seed-borne sweet corn pathogens with common names of the diseases they cause

Pathogens	Common names
Acremonium strictum W. Gans, syn. *Cephalosporium acremonium* Corda	Kernal rot
Cephalosporium maydis Samra, Sabet and Hingorani	Late wilt, late blight, slow wilt
Cochliobolus carbonum R. R. Nelson, syns. *Drechslera zeicola* (Stout) Subram and Jain, *Helminthosporium carbonum* Ullstrup	Charred ear mould, southern leaf spot
Cochliobolus heterostrophus (Drechsl.) Drechsl., syns. *Drechslera maydis* (Nisik.) Subram and Jain *Helminthosporium maydis* Nisik	Southern leaf spot or blight
Diplodia spp.	Dry ear rot, stalk rot, seedling blight, root rot, white ear rot
Gibberella fujikuroi (Saw.) Wollenw., syn. *Fusarium moniliforme* Sheld. G. f. var. *subglutinans* Edw., syn. *F. m. subglutinans* Wollenw. and Reink.	*Gibberella* ear rot, kernal rot, stalk rot, seedling blight
G. *zeae* (Schw.) Petch. syn. *Fusarium graminearum* Schwabe	Seedling blight, cob rot
Marasmius graminum (Lib.) Berk.	Seedling and foot rot
Sclerophthora macrospora (Sacc.) Thirum, Shaw and Naras., syn. *Sclerospora macrospora* Sacc.	Crazy top

Pathogens	Common names
Sclerospora philippinensis Weston, syn. *S. indica* Butler	Philippine downy mildew, crazy top
Ustilaginoidea virens (Cooke) Tak.	False smut, green smut
Ustilago maydis (DC.) Corda, syn. *U. zeae* (Schw.) Unger	Smut, blister-smut, loose-smut.
Erwinia stewartii (E. F. Smith) Dye, syns. *Bacterium stewartii* E. F. Smith, *Xanthomonas stewartii* (E. F. Smith), Dowson	Bacterial wilt, Bacterial leaf blight, Stewart's disease, white bacteriosis
Viruses	Maize leaf spot virus, Maize mosaic virus Sugar cane mosaic virus Wheat streak mosaic virus Corn stunt

REFERENCES

MacKay, D. C. and Eaves, C. A. (1962) The influence of irrigation treatments on yields and on fertilizer utilization by sweetcorn and snap beans, *Canad. J. Plant Sci.* **42** (2) 219–27.

Purseglove, J. W. (1972) *Tropical Crops, Monocotyledons.* Longman, London.

FURTHER READING

Airy, J. M., Tatum, L. A. and Sorenson, J. W. (1961) Producing seed of hybrid corn and grain sorghum. *In Seeds, Yearbook of Agriculture, 1961*, A. Stefferud (ed.). USDA.

Cordner, H. B. (1942) The influence of irrigation water on the yield and quality of sweet corn and tomatoes with special reference to the time and number of applications, *Proc. Amer. Soc. Hort. Sci.* **40**, 475–82.

Curtis, D. L. (1980) Some aspects of *Zea mays* L. (corn) seed production. In *Seed Production*, P. D. Hebblethwaite (ed.). Butterworths, London and Boston.

Feistritzer, W. P. (ed.) (1982) *Technical Guideline for Maize Seed Technology.* FAO, Rome.

Jugenheimer, R. W. (1976) *Corn Improvement, Seed Production and Uses.* Wiley, New York and London.

16 AMARANTHACEAE AND MALVACEAE

AMARANTHACEAE

Amaranthus spp.: African spinach, amaranth, bayam, bush greens, Chinese spinach

Celosia argentea L.: Cock's comb, green or white soko

These two genera are very similar in their morphology and cultural requirements and are therefore considered together. Important differences are given where appropriate.

According to Grubben (1977) the three *Amaranthus* species cultivated mainly as vegetables are *Amaranthus cruentus* (L.) Sauer, which is grown predominantly in Africa, *Amaranthus tricolor* L. the main Asian species and *Amaranthus dubius* Mart. ex Thell which is grown in the Caribbean. The *Amaranthus* species which are cultivated for the production of flour are usually referred to as the 'grain amaranths'. *Amaranthus hypochondriacus* L. is the most important species cultivated for this purpose although the vegetable amaranth *A. cruentus* L. is also used for grain. The evolution and history of the grain amaranths was reviewed by Sauer (1976).

Amaranthus and *Celosia* are often considered as the most important green leafy vegetables of the tropics because they provide minerals and vitamins (especially vitamin A) in the diet of many people in developing countries (e.g. in the Sahel area of Africa, Fig. 16.1).

Cultivar descriptions of *Amaranthus* and *Celosia*

Genus and species:

Uses: suitability for specific locations and seasons

Yield of leaf and/or grain per unit area and percentage of dry matter content.

General morphology and plant form.
  Apical dominance, degree of branching

Leaf   Number from time of sowing or transplanting, shape and pigmentation

Inflorescence   Time to flowering and specific daylength response
                Colour
                Form

Fig. 16.1   Seed production of *Amaranthus cruentus*, Benin (by courtesy of Dr
G. J. H. Grubben).

General characteristics

Resistance to specific pathogens, e.g. *Choanephora cucurbitarum* (Berk. et Rav.)
Thaxter and *Meloidogyne* sp.

Resistance to drought

Agronomy

Soil, pH and nutrition

There are no specific references in the literature to the optimum soil pH for
Amaranthus or *Celosia* but analysis of soils in Dahomey by Schelhaas (1974)
indicated that the crops were frequently grown on soils ranging from pH 4.6
to 7.5. Although Grubben (1976) included liming as a variable in his investigations into the use of town refuse and NPK applications on amaranth production, there was no effect on yields from liming; however it is possible that
the quantities of lime applied were not sufficient to increase the soil pH.
Other results of Grubben's investigations relating to pH indicate that these
genera may exhibit lime-induced chlorosis and that while both crops tolerate
a wide pH range, soils with a pH higher than 7.0 should not be used.

   Grubben (1976) reported that the best response to fertilizers was obtained
when the N : P : K ratios were 1 : 1 : 2. The response to potassium was especially noticeable and the plants readily exhibited potassium deficiency
symptoms on the poorer soils.

Sowing and young plant management

It is customary to broadcast the seed in seed beds at the rate of 2 g per square metre for *Amaranthus* and 3 g per square metre for *Celosia*. These sowing rates are sufficient to produce 1000 seedlings ready for transplanting after approximately three weeks (Grubben 1976). The seedlings are planted out in rows 60–80 cm apart with 40–50 cm between the plants. The sowing rate for both crops is 1 kg/ha when drilled direct.

It is customary to pinch out the growing point of those cultivars which have a relatively small apical inflorescence four weeks after planting to encourage development of secondary shoots. This is especially important with the cultivars which have strong apical dominance, but the growing points are not removed from cultivars with a relatively large apical inflorescence.

Flowering and pollination

There are neutral and short-day types in both genera which flower in all the tropical areas where the crop is cultivated.

Amaranthus is wind pollinated and *Celosia* is predominantly insect pollinated (Grubben 1976).

Isolation

Plots for seed production should be isolated from all sources of pollen contamination by a minimum of 500 m. There are wild forms of both *Amaranthus* and *Celosia* which are common weeds in the tropics and it is therefore important to ensure that the seed production plots and adjacent areas are kept weed-free.

Roguing stages

1. *At planting out*   Check that plants are true to type, and that early flowering plants are removed; appropriate vegetative characteristics, including leaf number per plant, leaf shape and size according to plant stage are confirmed; the degree of leaf pigmentation is confirmed according to cultivar.

2. *Before flowering*   As above, in addition check that plants are of appropriate height and that the degree of branching is according to the cultivar.

3. *At start of flowering*   Appropriate height and general morphology. Inflorescence and flower colour.

Harvesting

The method of harvesting depends on the scale of production and to some extent on the cultivar. Plants of cultivars with an apical inflorescence are cut when the seeds are mature and there is a general yellowing of the plant. Seed heads are cut from the cultivars with axillary inflorescences when they mature and up to four successive harvests are made to obtain the maximum potential seed yield.

Seed yield and 1000 grain weight

Seed yields of up to 2000 kg/ha for *Amaranthus* and 600 kg/ha for *Celosia* have been reported by Grubben (1976). The 1000 grain weight of *Amaranthus* is 0.3 g and for *Celosia* is 1 g.

Seed-borne pathogens

The two main seed-borne pathogens of *Amaranthus* are *Alternaria amaranthi* (Peck) Van Hook (blight) and Strawberry latent ringspot virus.

MALVACEAE

Hibiscus esculentus L., syn.
Abelmoschus esculentus (L.) Moench: Okra, Lady's finger, gombo

Okra is an important vegetable in the tropics and sub-tropics. It probably originated in the Ethiopian region of Africa, but is now widely grown in Africa, especially in the Sudan, Egypt and Nigeria; it is also very important in other tropical areas including Asia, Central and South America. In recent years there has been interest in growing it as a protected crop in heated greenhouses in Northern Europe.

Okra is cultivated mainly for its 'pods' which are cooked and eaten, although in some countries the young leaves are cooked also. There are many local cultivars, often with very specific characteristics, e.g. the high mucilaginous content preferred in the Sudan. The modern cultivars produced in the USA, e.g. 'Clemson Spineless', fruit relatively early on more compact plants than the traditional African types.

Cultivar description of okra

General use of the cultivar, early production, canning, drying or for fresh market

General character of plant, degree of branching and plant height, season, relative yield, resistance to drought, duration of cropping, reaction to rainy season, response to daylength.

Pigmentation of stem and leaf
(NB   Leaf lobing tends to be more prominent higher up the plant.)

Flower   density of yellow colour, relative size and extent of pigment at
           bases of petals (Fig. 16.2)

Pod   relative length, shape and 'beak' at end of fruit (Fig. 16.2)
           colour – white, green, yellow-green, yellow or red
           degree of spininess
           transverse section, circular or angular
           mucilage content
           fibre content

Resistance to specific pathogens, e.g. mosaic virus

Fig. 16.2   Two distinct local cultivars of okra grown in the Sudan.

Agronomy

Okra is grown as a warm season crop throughout the tropics and although
the market crop can be produced in the rainy season, the seed crop is best
timed so that pod ripening occurs under relatively dry conditions. Seed ripening occurs approximately one month after the stage for normal harvesting
of the edible crop.

The crop is grown as an annual although there are a few types grown as
perennials which originate mainly from Ghana.

Nutrition

Okra is slightly tolerant of acid conditions and can be grown in soils with a pH between 6.0 and 6.8. The general fertilizer recommended is a N : P : K ratio of 1 : 2 : 1 applied during the final stages of preparation. Nitrogenous fertilizers are also applied as top-dressings at a rate of up to 50 kg/ha depending on the leaching rate early in the growing season.

Flowering

The majority of okra cultivars are short-day plants requiring a minimum day temperature of 25 °C, but thriving in temperatures up to about 40 °C, although there are reports of exceptions to this. It is generally believed that the best seed crop is produced in areas with a relatively small day and night temperature differential.

In some tropical areas the growing points are 'pinched' out or stopped in order to increase the flower number on the extended laterals, but this is not practised in large-scale production.

Pollination

The flowers are insect pollinated although self-fertile cross-pollination occurs.

Isolation

Recommended isolation distances vary from one country to another, but a minumum isolation distance of 500 m is desirable.

Okra roguing stages and main characteristics to be observed

1. *Before flowering*   Check the general plant height and habit, pigmentation of leaves, petioles and stems; remove plants with virus symptoms.

2. *Flowering*   Check the relative size and colour intensity of flowers; remove plants with virus symptoms.

3. *Fruiting*   Check that fruit is true to type; remove plants with virus symptoms.

Sowing

A sowing rate of 8 kg/ha is used for large-scale production of the compact cultivars on the flat in rows 45 cm apart. In areas where furrow irrigation systems are used the sowing rate is 5–6 kg/ha with 90 cm between the rows. Plants are thinned to 15–30 cm apart in the rows according to the vigour of the cultivar. Some seed producers sow in blocks of 10–15 rows with a gap of two rows width between blocks to facilitate field inspections.

Harvesting and seed extraction

There is a sequential ripening of okra pods on a plant. The fruits of the angular fruited types have a tendency to split when the seed ripens (Fig. 16.3). The maturity of seeds is associated with the pods becoming grey or brown according to the cultivar.

The traditional hand harvesting of ripe pods is still done in many tropical areas where there is adequate labour, although the crop is combined in the USA.

Seeds are extracted after the hand-harvested pods become dry and brittle. The most efficient method of hand seed-extraction is to twist the pods open. Alternatively the pods are either flailed or the seeds are extracted with a stationary thresher.

In some areas of the world, especially Malaysia, the initial pods are harvested as a fresh vegetable and the later pods are retained on the plants for seed production (Soo 1977). This practice probably does not reduce the potential seed yield in areas with a sufficiently long growing season, as removal of early pods tends to encourage further extension growth and flower development.

Seed yield and 1000 grain weight

A relatively high seed yield of 1500 kg/ha is achieved in the okra seed producing areas of the USA but in many tropical countries the yield rarely exceeds 500 kg/ha.

The 1000 grain weight of okra seeds is *c.* 50 g.

The main seed-borne okra pathogens with common names of the diseases they cause

Pathogens	Common names
Ascochyta abelmoschi Harter	*Ascochyta* blight
	pod–spot
Choanephora cucurbitarum (Berk. and Rav.) Thaxter	Okra fruit rot
Fusarium solani (Mart.) Sacc.	
Glomerella cingulata (Stonem.) Spauld and Schrenk. syn. *Gloeosporium cingulatum* Atk.	
Rhizoctonia solani Kühn	
Viruses	Okra leaf curl
	Mosaic

Fig. 16.3   Sequential seed pod ripening and shattering of okra.

REFERENCES

Grubben, G. J. H. (1976) *The Cultivation of Amaranth as a Tropical Leaf Vegetable.* Royal Tropical Institute, Amsterdam.

Grubben, G. J. H. (1977) *Tropical Vegetables and Their Genetic Resources*, H. D. Tindall and J. T. Williams (eds). International Board for Plant Genetic Resources, IBPGR/FAO, Rome.

Sauer, J. D. (1976) Grain amaranths. In *Evolution of Crop Plants*, N.W. Simmonds (ed.). Longman, London and New York.

Schelhaas, R. M. (1974) *De Tuinbouwgronden van Zuidoost-Dahomeg. (Les sols des cultures maraichères du Sed-Est Dahomey.* Department Agricultural Research, Royal Tropical Institute, Amsterdam.

Soo, T. T. (1977) Local horticultural seed production. In *Seed Technology in the Tropics*, H. F. Chin, I. C. Enoch and R. M. R. Harun (eds). University Pertanian Malaysia.

FURTHER READING

Martin, F. W. and Ruberté, R. (1978) *Vegetables for the Hot, Humid Tropics, Part 2. Okra.* Mayagüez Institute of Tropical Agriculture, Puerto Rica.

Martin, F. W. and Telek, L. (1979) *Vegetables for the Hot, Humid Tropics, Part 6. Amaranthus and Celosia.* USDA, New Orleans.

Oomen, H. A. P. C. and Grubben, G. J. H. (1978) *Tropical Leaf Vegetables in Human Nutrition.* Royal Tropical Institute Amsterdam and Orphan Publishing Company, Willemstad, Curacao.

APPENDIX 1

METRIC (SI) CONVERSIONS

Distance

1 millimetre (mm) = 0.039 inches
1 centimetre (cm) = 0.393 inches
1 metre (m)   = 3.28 feet or 1.09 yards
1 kilometre (km)   = 0.62 miles
1 inch   = 25.40 mm or 2.54 cm
1 foot   = 30.48 cm
1 yard   = 0.91 m
1 mile   = 1.61 km

Area

1 hectare (ha)   = 2.47 acres
1 square kilometre (km^2) = 247 acres
1 acre   = 0.40 ha
1 square mile   = 258.9 ha = 2.59 km^2

Capacity

1 litre (l) = 0.22 Imperial gallons
        = 0.26 US gallons
1 Imperial gallon   = 4.55 l
1 US gallon   = 3.78 l

Weight

1 gram (g)   = 0.035 ounces
1 kilogram (kg) = 2.20 pounds
1 tonne (t)   = 1.10 short tons
1 ounce   = 28.35 g
1 pound   = 453.59 g
1 short ton = 2000 pounds = 0.907 t

BOTANICAL INDEX

GENERAL INDEX